SET PHASERS ON STUN

And Other True Tales of Design, Technology, and Human Error

SECOND EDITION

STEVEN CASEY

Aegean

Aegean Publishing Company
Santa Barbara

Published in the U.S.A. by
Aegean Publishing Company
Post Office Box 6790, Santa Barbara, California 93160

SET PHASERS ON STUN
AND OTHER TRUE TALES
OF DESIGN, TECHNOLOGY, AND HUMAN ERROR

ISBN 0-9636178-8-5 (hc)
978-0-9636178-8-0
If not available from your local bookstore,
this book may be ordered directly from the publisher.
Send the cover price plus $4.50 for shipping and handling costs to the above
address. California residents add applicable sales tax.

Publishers Cataloging in Publication Data
Casey, Steven M.
Set phasers on stun
and other true tales of design, technology, and human error.
Second Edition
p. cm.
Includes bibliographical references
ISBN 978-0-9636178-8-0 (hc) : $29.00
1. Technology. 2. Human engineering.
3. Human factors 4. Engineering design. 5. Title
TA 166 C37 1998 620.8'2 97-77875

TABLE OF CONTENTS

For he, thy choice flower stealing,
the bright glory
Of fire that all arts spring from, hath bestowed it
On mortal men.
And so for fault like this
He now must pay the Gods due penalty...

PROMETHUS BOUND
Aeschylus

PROLOGUE

Since the publication of the first edition, *'Set Phasers on Stun' and Other True Tales of Design, Technology, and Human Error* has achieved a certain level of notoriety as a stirring, unique, and disturbing trek into the real world of human factors and technology. Intentionally void of esoteric cognitive models which seek to define everything but end up explaining nothing, *Set Phasers on Stun* was written to take the reader on a different course, one in which the perceptions, quandaries, and actions of real people interacting unsuccessfully with real technology could be appreciated and understood without having to suffer the same unfortunate consequences. By focusing on the experiences of the individual within the context of a specific setting, it was and is my hope to convey how *incompatibilities between the way things are designed and the way people perceive, think, and act* can result in human error, or, more accurately, design-induced human error.

In this, the second edition, I have added two new stories to the original eighteen, for a total of twenty. *The Price of the Amagasaki,* Chapter 19, relates the tale of a British Navy diver and his efforts to sink a Japanese ship using a pure oxygen rebreather at the height of the Second World War. The diver was one of many who suffered the grave consequences of using a technology that was not well understood, yet put into service due to the pressures of war and the tactical advantage afforded by a device providing unparalleled stealth. Designers and users make trade-offs, and the outcome of the story is clearly influenced by the wartime context in which the system was engineered and used. The story is also an example of technology getting a step ahead of knowledge of some basic human factors.

Chapter 20, *Murphy's Law and Newton's Law,* is a contemporary tale about the design of a small test kit and a catastrophic event at ESRANGE, the European space range in northern Sweden. Newton's third law of motion defines the equivalence of physical actions and reactions, the basic physics of rocket flight. 'Murphy's law,' brought to us courtesy of Captain Edward A. Murphy of the U.S. Air Force, predicts the uncertainties of systems: "If anything can go wrong, it will." The former describes the deterministic nature of physics and physical reactions (at least as they are experienced in everyday life); the latter addresses the probabilistic nature of human behavior. A good designer should have anticipated the actions of the technicians at ESRANGE and avoided the design that resulted in the seemingly unimaginable event.

Set Phasers on Stun has been adopted as a supplemental reading text in courses at many colleges, universities, and companies, something I did not anticipate but I do find gratifying. Of particular interest has been the breadth of the courses in which the book has been used as a text - - human factors and accident analysis classes, courses in industrial design, ergonomics, computer science, technology and society, systems safety, psychology, industrial engineering, electrical engineering, management science, and even English. A Japanese version has been published by Kagaku-Dojin Publishing Company of Kyoto. The translation was performed by Motoyuki Akamatsu of the Japanese National Institute of Bioscience and Human Technology. Kagaku-Dojin and Motoyuki Akamatsu changed the title for the Japanese language edition. In Japan, the very American '*Set Phasers on Stun*' is '*Human Error - - This is How Accidents Happened.*'

A few readers have asked why I have not analyzed the stories, broken from the narrative and point of view, and explained, item by item and point by point, what should have been done differently and what lesson is to be learned. My response to this is simple: It is the driver of the car, not the passenger in the back seat, who learns his way. The facts have been presented, the clues are all there. Take the perspective of the user and find the answers on your own. I, for one, think that they are reasonably obvious.

These stories are fundamentally about the design of things used by people. The intention of the book is emphatically not to be sensationalistic or to satisfy my own or someone else's morbid curiosity about technological disasters. Rather, I have written the stories to serve as vehicles for thinking about and understanding what happens when designers of products, systems, and services fail to account for the characteristics and capabilities of people and the vagaries of human behavior. The human component of a system is every bit as important as any nut, bolt, circuit, or line of software code. By reading these stories, I hope that designers, engineers, programmers, managers, and human interface designers who shape modern technological devices and systems are better able to anticipate human behavior and design accordingly, or, at the very least, recognize the growing need to incorporate human engineering analyses, methods, and principles into design. Society has become highly dependent on interconnected and tightly linked technologies, technologies in which the actions of a few influence the lives of many, technologies that can bring countless benefits but can also amplify the consequences of human error in ways that were never before possible or never anticipated. Traditional engineering knowledge - - of electronics, chemistry, physics, structures, and materials - - is insufficient in and of itself for the design of technologies which play such a profound role in our lives. The future will require more of designers and their designs as systems become more complex, more intertwined, and even more, not less, dependent on human capabilities and limitations.

I would like to thank the many dozens of individuals who assisted me with the first and second editions by providing useful leads and story ideas. I am also grateful to the many readers who have sent me material on other promising stories; it is my intention to put this information to good use in follow-on volumes.

SET PHASERS ON STUN

Voyne Ray Cox, 33, just "Ray" to his family and numerous friends, winced a little as the bare skin of his stomach and chest pressed down on the icy tabletop positioned beneath the massive Therac-25 cancer radiation therapy machine. The table top wasn't really all that cold, it just seemed that way due to the unseasonably warm weather and the constant air conditioning inside the East Texas Cancer Center here in Tyler. It was Ray's ninth radiation treatment since the surgery for the removal of the tumor on his left shoulder, and he expected the session to be just as uneventful as the previous eight. He should be out of there in no time at all.

Mary Beth, the radiotherapy technician, was being every bit as pleasant as she had been during his other visits. She commenced with the customary greetings and banter and helped him get positioned on the perfect spot on the table. The surgery and radiation treatments had proved to be better than he had feared, and he was looking forward to getting it all behind him and returning to work in the oil fields east of town. Not one to be fazed by adversity, Ray knew what he had to do, and he was determined to get on with the treatment and get this cancer thing well into the past.

Ray, lying chest-down with the side of his face resting on the table, listened to the familiar instructions about remaining still. He watched Mary Beth from his prone position as she operated the hand-held control console that rotated the table and him to the proper position underneath the machine's gantry. She told him to stay still and then walked out the exit of the treatment room to the small control room outside down the hall. Ray raised his head a little and looked up at the imposing device poised overhead. To Ray, the radiotherapy machine was just a big and attractive-looking piece of gear, all packaged-up nicely in clean and seamless sheets of plastic and metal. But, as it had been explained to him by the physicians and technicians at the center, the million dollar Therac-25 was the state-of-the-art in cancer treatment equipment. Beneath the sleek skin was a complex and downright ominous-looking machine, capable of delivering a beam of high-energy radiation to any point on or in a person's body. There were two keys to treatment success: hitting the cancer cells with pinpoint accuracy and having many separate treatments of relatively lower doses rather than a single treatment of one large dose. Ray Cox knew that he would be back in the treatment room many times over the next few weeks, and any remaining cancer cells from the small tumor that had been removed from his back would eventually be killed. By now, Mary Beth was in the control room, so he put his head back down after deciding it best to just mind his own business and take the opportunity to relax and be still.

Inside the small control room, Mary Beth began to enter the commands into the computer keyboard to initiate the treatment. She was working with a common Digital Equipment Corporation VT100 terminal which in turn was connected to a PDP-11 computer that controlled the radiotherapy accelerator. The control system would aim the accelerator with pinpoint accuracy and, when ready, briefly fire a radiation beam of prescribed intensity. Ray Cox should not feel a thing.

There was a video camera inside the treatment room, but there was no picture on Mary Beth's television screen in the control room (the video monitor was not plugged in). The voice intercom between the two rooms was also inoperative. Neither of these things were viewed as being particularly significant, as Ray Cox had gone through this before and was now lying on the table right where he should.

The Therac-25 had two modes of operation, something that made it unique in the radiotherapy marketplace. One mode was a high-power "x-ray" mode utilizing the full 25 million electron volt capacity of the machine. It was selected by typing an "x" on the keyboard. This put the machine on maximum power and automatically inserted a thick metal plate just beneath the beam. When passed through the metal plate, the beam was transformed into an x-ray which was used to radiate tumors inside the body. The plate also lowered the intensity of the beam.

The other setting was a relatively low-power "electron beam" mode and was selected by pressing the "e" key. Ray Cox was scheduled to be treated with the "electron beam" mode. He would receive a painless burst of about 200 rads to the spot on his shoulder.

Mary Beth pressed the "x" key, moved on to the next entry on the keyboard, and realized suddenly that she had entered the wrong letter. She meant to enter an "e" to set the machine for "electron beam" mode but had mistakenly entered an "x" and set the machine on "x-ray" mode. It was a simple enough slip, and one that certainly had no consequence since the treatment had not yet begun. Not one to waste time, Mary Beth quickly pressed the "up" arrow key to select the "edit" functions from the computer display. This enabled her to change the incorrect "x-ray" setting to the correct "electron beam" setting, which she did by pressing the "e" key. The screen now indicated to her that she was in the "electron beam" mode. The error corrected, she quickly pressed the return key on the keyboard to move the cursor to the bottom of the screen to wait for the "beam ready" display indicating that the machine was fully prepared to fire the narrow beam of radiation down onto Ray Cox's back. All of this took place within the span of eight seconds.

She had no idea that no one had ever entered this unusual but not at all unexpected sequence of commands in the thousands of times this particular Therac-25 had been operated. And the engineers for the Canadian manufacturer of the Therac-25, Atomic Energy of Canada, Ltd. (AECL), had not considered it possible for a technician to enter this sequence of commands in less than eight seconds. Accordingly, AECL did not test this unique sequence of inputs during the machine's development a few years before.

Mary Beth thought nothing of her small error or the quick steps she had taken to correct it. But unknown to her, AECL, and Ray Cox, the stage was set for disaster. Her rapid and unique series of inputs had tripped-up the computer. It retracted the thick metal plate used during x-ray mode - - but left the power setting on maximum. Her computer screen showed that the machine was in the necessary "electron beam" mode, but it was actually now in a debased operating setting and poised to deliver a blast of 25,000 rads down onto Ray Cox's back - - in a proton beam powered by 25 million electron volts!

She looked back down at the computer screen just as the "beam ready" command appeared telling her that the machine was primed and ready to fire. Well-trained in the procedures, she then pressed the "b" key to turn the beam on...

Ray Cox saw a flash of blue light, heard a frying sound, and felt the invisible lightning bolt of high-energy radiation shoot down from the Therac-25 into his back. It was as if a red-hot fireplace poker had been jammed through to his chest. He jolted reflexively. The pain was excruciating, nothing remotely like the other treatments.

Inside the isolated control room, Mary Beth's computer screen simultaneously displayed "Malfunction 54," indicating that something was not working and that the treatment had not been initiated. Having received no feedback that the machine had fired, she quickly reset the Therac-25 so that she could try it again...

Out on the table, Ray was rolling onto his side, his shoulder feeling on fire. The blue light flashed again, there was a sizzling sound, and Ray, now conditioned and forewarned, began to scream out for it to stop. But, before he could expel the complete force of his terrified shout, the proton beam shot down from above, this time into his neck. His chest muscles constricted, squeezing the air out of his lungs. The pain was as intense as anything he had ever felt, and for a moment he thought he might pass out. He slowly caught his breath and held it in his chest while trying to maintain some level of control.

A few moments later he took a deep cleansing breath, joggled his head once, and called out to Mary Beth.

"Hey, are you pushing the wrong button?"

At nearly the same moment, Mary Beth's computer screen in the control room showed that the Therac-25 was primed and ready once again, and she entered the "b" to fire the high-energy beam. Inexplicably, the machine responded by displaying "Malfunction 54." She had never had any problems with the machine before, and certainly had no idea what was meant by the error code...

Outside on the treatment table, Ray Cox was hit for a third time with another deadly and invisible shot to his shoulders and neck. He might just as well have been standing at the wrong end of a firing range. Fearing for his life and writhing in pain, he jumped from the table and ran to the door, where he bumped into technicians walking down the hall.

Mary Beth eventually came out of the control room and met Ray at the nurses' station, where he explained to her that he had received repeated and painful "electric shocks" while lying on the table. She responded by saying how strange that was and that nothing like that had ever occurred before. She had no idea what might have happened, but that there was no need to be concerned because the machine had malfunctioned and shut down automatically. According to the display, Ray had received only one tenth of his prescribed treatment dose.

Faced with an inoperative machine and a disrupted schedule, Mary Beth called Fritz Hager, the radiological physicist in charge of the Therac-25 at the center, and asked him to come down to examine the equipment. Lee Schlichtemeier, Ray's radiological oncologist, eventually made his way to the treatment room as well. They found nothing physically wrong with Ray or the machine, but decided to call AECL and discuss the event. After conducting some recommended tests of the equipment and finding nothing that suggested anything remotely abnormal, they continued use of the machine that same afternoon. There were, after all, other patients in the queue.

Just three weeks later it happened again with another patient, a 66-year old man by the name of Verdon Kidd, undergoing treatment for a growth on his ear. Mary Beth inadvertently typed an "x" instead of an "e" and then corrected her error by entering the "edit" routine and changing the mode to "e." The man was severely "shocked" about eight seconds later. Mary Beth watched in amazement from the control booth as the Therac-25 shut down, apparently without administering the planned treatment. The computer screen registered a "Malfunction 54."

With the second occurrence it became apparent to Fritz Hager that there were serious problems with the control system, problems that occurred when the "edit" function was used to quickly change the setting from "x-ray" mode to "electron beam" mode. Although the machine told the operator it was operating in the "electron beam" mode, it was actually operating in a hybrid proton beam mode and delivering blasts of 25,000 rads with 25 million electron volts - - more than 125 times the prescribed dose. And to make matters worse, Mary Beth had delivered multiple "treatments" to Ray Cox because of the lack of feedback and her conclusion that no treatment had been administered. The blue flash Ray saw before each blast was from Cherenkov radiation, a rare phenomenon seen only when a stream of electrons is accelerated to an extreme velocity.

Hager immediately called AECL, as well as other U.S. users of the Therac-25, and alerted them to the problem. Subsequent investigations lead to the discovery of similar overdoses in clinics in Marietta, Georgia; Ontario, Canada; and Yakima, Washington.

Around the same time, Ray Cox's doctors began to suspect that he had been subjected to a massive radiation overdose. He was starting to spit up blood, and terrible radiation burns were appearing on his back, shoulder, and neck. Over the next few months the tissues hit by the beams died and sloughed off, leaving massive, gaping lesions in his upper body. Before his death four months later, Ray Cox maintained his good nature and humor, often joking in his east Texas drawl that "Captain Kirk forgot to put the machine on stun."

REFERENCES AND NOTES

A computer glitch turns miracle machine into monster for three cancer patients (1986). *People Weekly*, 26 (November 24), 48-50.

Fatal radiation dose in therapy attributed to computer mistake (1986). *New York Times*, June 21, 50.

Jacky, Jonathan (1990). Risks in medical electronics. *Communications of the ACM*, 33 (12), 138.

Joyce, E. (1986). Firm warns of another Therac-20 problem. *American Medical News*, November 7, 20-21.

Joyce, E. (1986). Malfunction 54: unraveling deadly medical mystery of computerized accelerator gone awry. *American Medical News*, October 3, 1 (8 pages).

Joyce, E. (1986). Software bug discovered in second linear accelerator. *American Medical News,* November 7, 20-21.

Joyce, E. (1987). Software bugs: a matter of life and liability. *Datamation,* May 15, 33 (10), 88-92.

Lee, L. (1992). Computers out of control. *BYTE,* February, 344.

Lee, L. (1992). *The day the phones stopped: how people get hurt when computers go wrong.* New York: Donald I. Fine, Inc.

Moreno, M. (1989). Product recalls. *The Boston Globe,* April 10, 14.

Neumann, P. G. (1987). Illustrative risks to the public in the use of computer systems and related technology: update on Therac-25. *ACM SIGSOFT Software Engineering Notes,* 12 (3), 7.

Parnas, D. L., van Schouwen, J., and Kwan, S. P. (1990). Evaluation of safety-critical software. *Communications of the ACM,* 33 (6), 636-648.

Saltos, R. (1986). Man killed by accident with medical radiation. *The Boston Globe,* June 20, 1.

Therac-25: optimal high energy linac produces a true 25 MV photon beam (undated Therac-25 sales literature). Atomic Energy of Canada Limited, Medical Products: Ottawa, Canada.

Therac-25 linear accelerator: specification no. GS3500 (undated). Atomic Energy of Canada Limited, Medical Products: Ottawa, Canada.

Thompson, R. C. (1987). Faulty therapy machines cause radiation overdoses. *FDA Consumer*, 21 (10), 37-38.

Waldrop, M. M. (1989). Congress finds bugs in the software. *Science*, November 12, 753.

Young, F. E. (1987). Validation of medical software: present policy of the Food and Drug Administration. *Annals of Internal Medicine*, 106, 628-629.

A pseudonym has been used for the radiotherapy technician. Any similarity to her real name is purely coincidental. All other names and events are based on published accounts and are believed to be accurate.

RETURN FROM SALYUT

On the evening of June 29, 1971, Viktor Patsayev, Georgi Dobrovolsky, and Vadim Volkov were busy shutting down the last of the control systems aboard Salyut and retreating to the Soyuz 11 command module for their return home. Three more cosmonauts were scheduled to return to Salyut in a matter of a few weeks, so they wrote a brief note to welcome them to the space station. Georgi stuck it to the wall of the transfer/docking module, and Viktor laughed to himself as he envisioned the new space crew reading it as they entered Salyut. Viktor thought about how much the past three weeks had enriched his own life, and he felt a little smug leaving a few words of wisdom for the future residents of Salyut. Viktor and his two comrades, Georgi Dobrovolsky, the mission commander, and Vadim Volkov, the flight engineer, were now about to complete a three-week stay in Salyut, the world's first space station. The mission had been a resounding success, and Viktor was prepared to return to a hero's welcome on Russian soil.

At 9:15 p.m. the mission controller at the manned space flight control center in Kalinin outside Moscow instructed Vadim to close the hatch between the Soyuz orbital module and the Salyut transfer/docking module for the last time. Viktor, already sitting in his position in the cramped Soyuz module, looked up to see the soles of Vadim's shoes as he worked in the transfer/docking module above his head. Vadim rechecked the hatch, and all indicator lights signaled that they had proper closure. He floated down and maneuvered himself into his seat to Georgi's right. Vadim reached up and closed the hatch between the command module and the orbital module, checked the seal, checked it once again, and then sat down and buckled himself in. Georgi informed ground personnel that they had closed the hatches and all systems looked good. Viktor, seated on Georgi's left, felt Georgi's left shoulder rubbing against him every time either one of them moved. He reminded himself that he would have to sit there for only a few hours.

After many minutes of final preparation, the command module, firmly attached to the orbital module, disengaged from Salyut. Soyuz 11, like earlier Soyuz and Vostok spacecraft that preceded its development, was highly automatic. The cosmonauts interacted very little with the onboard systems except when docking and undocking. For the most part Viktor, Georgi, and Vadim monitored the displays on the austere control panel in front of them and observed Salyut as they pulled slowly away. Salyut was a striking sight. Viktor admired her green insect-like form through the port window to his left. The large craft hung suspended against the jet black curtain of space, her skin brightly illuminated by the unfiltered rays of the sun.

Soyuz flew adjacent to Salyut for the next four hours as they approached their planned reentry point. Viktor made use of the time by stowing some additional materials and double-checking the reentry procedures and systems. Although the cabin was cramped, it would have been far worse if they were wearing their space suits. Previous Soyuz cosmonauts had worn the characteristically white Soviet space suits - - but they were one- and two-man flights. Soyuz was basically a two-man capsule, and it had been necessary to exclude the suits from the three-man configuration for the simple reason that the men, their suits, and the support equipment did not fit inside the cabin. After all, the American Apollo capsules held a crew of three, so they could certainly do the same. All they had to do was eliminate the space suits which were not really essential anyway. Besides, their wool jump suits were far more comfortable.

Viktor Patsayev was well qualified for the tasks assigned to him. Like Vadim, he was a civilian who had longed to fly in space. He held a masters degree from the Penze Industrial Institute and had been a design engineer at the Central Aerological Observatory. Viktor previously demonstrated his ability to perform in space on Soyuz 7. Now he was flying as the Soyuz 11 Test Engineer. He was known by his comrades in the cosmonaut corps as being a bit on the quiet side, but was bright, highly competent, and always dependable.

Georgi Dobrovolsky, the commander of the mission, was a Lieutenant Colonel in the Soviet Air Force. Like other Soviet space shots this one was commanded by a military man. Vadim considered Georgi to be the most popular cosmonaut in the entire corps. He had a reputation not only as a skilled pilot, but as a man who enjoyed good company and conversation. He was a perfect candidate for the first long-duration space flight aboard a space station. The same could be said for Vladislav "Vadim" Volkov, the flight engineer. Although he was closer in age to Viktor, Vadim's personality was very much like that of Georgi - - gregarious, personable, and entertaining. Like Georgi, this was his first trip into space.

Viktor gazed out the small left-side porthole of the Soyuz command module for over a minute. The eastern coast of the Soviet Union and the Sea of Japan slid by below. Like Georgi and Vadim, viewing earth through one of the few small windows on the space station was his favorite pastime, but he rarely had the opportunity because of other, more pressing duties. He had no particular thoughts at the moment, the isolation of the past 23 days had taken its toll on his ability to concentrate during moments like this. It was such a pleasure to simply stare out the window and absorb the view. The top-down perspective of the blue water, coastal islands, and cumulus clouds was something so few people had ever seen, and it was just as absorbing now as it had been three weeks before.

The reentry burn began precisely at 1:35 a.m., Moscow time, after Soyuz obtained the proper reentry position over earth and oriented herself so the cosmonauts faced backwards, away from the direction of travel. The retro rockets on the command module ignited, sending a noisy shudder throughout the small spacecraft. The harsh shaking was disturbing, but Viktor's concern faded as the motion subsided into a forceful and prolonged rumble for the remaining seven minutes of the reentry burn. Viktor was privately relieved about the successful ignition - - even though they had not anticipated any difficulties with the scheduled reentry maneuver. He sat back in his seat, facing the emptiness of space, as Soyuz 11 dropped out of orbit and approached the upper layers of the atmosphere above the Soviet Union. The reentry burn slowed their fall, and Viktor became aware of the return of weight to his arms, something he had not experienced during his three-week stay in space. In less than 30 minutes they would decelerate from a speed of more than 27,000 kilometers per hour to a dead stop.

The flight controllers in Kalinin and the landing crews located 500 kilometers southwest of Sverdlousk made final preparations for the landing. Soyuz 11 should be nearing the atmosphere about now. The retro rockets would separate automatically from the command module, exposing the reentry shield just behind the backs of the three men. The forces on the cabin would increase to just under four gs, and flames from the heat of reentry would glow outside the portholes on the left and right sides of the small craft. Communication would not be possible due to the conditions of the fiery reentry.

The computers on board Soyuz, as programmed, shut down the retro rockets. The sudden silence within the cabin startled Viktor, but again, it was not an unexpected event. He noted the return of complete weightlessness as his arms lifted up away from his bent legs. The instrument assembly module, the command module, and the orbital module fell freely toward the atmosphere. Everything was going according to plan. They were no longer decelerating from the retro burn, just falling at thousands upon thousands of kilometers per hour through the black night. Seconds later the computers sent a signal to the explosive bolts connecting the command module with the orbital module. The bolts exploded, as instructed, and the command module fell away from the orbital module.

Water vapor and dust particles suddenly appeared out of nowhere, suspended in the cabin. A torturous pain spread across Viktor's face and forehead. The spacecraft began to spin unexpectedly. Flight computers, sensing that the craft was no longer oriented properly, instructed the positioning rockets to fire and keep the rear of the craft aimed toward the atmosphere below. Viktor, Georgi, and Vadim looked at the suddenly rotating attitude display on the console directly in front of them and realized that something had gone very, very wrong.

"The orienting rockets are firing!" Georgi shouted. "We're

rolling!"

Vadim, nearly at the same moment, shrieked, "We're losing pressure!"

Ground controllers on earth, including listeners at a Western tracking station eavesdropping on the flight, heard terse and unintelligible shouting from the craft, followed by a loss of radio telemetry. Loss of telemetry was normal during reentry, but usually not for another minute or so. Listeners on the ground weren't quite sure what, if anything, was going on.

After about 10 seconds of complete puzzlement and confusion, Viktor realized that the atmosphere within the command module was rapidly escaping into the vacuum outside. The force of the escaping air was acting just like an orienting jet on the capsule! The real orienting rockets must be firing to maintain the prescribed position of the spacecraft. But what was it? What was wrong? God, the hatch! It was his worst nightmare come true. Integrated into the center of the hatch that now separated the men from the vacuum outside was a pressure equalization valve, designed to open automatically when Soyuz entered the atmosphere and deployed her parachute. Firing of the explosive bolts during separation inexplicably caused the valve to open to the outside, enabling the cabin atmosphere, kept at sea level pressure, to blast through the opening into the vacuum.

Viktor pushed down with the full force of his legs and left hand, lunging toward the hatch above their heads with his right arm extended. The restraints dug deep into his right shoulder, holding him back in his seat. He clawed frantically at the latch on his restraining belts with his hands, finally getting his fingers on the buckle and pulling up to get some slack in the straps. With the straps loosened he pressed down again with his feet and right hand. Thrusting his torso upward, he grasped the hatch wheel with his left hand, pulled himself up as far as the restraints would allow, and grabbed a small handle connected to the open pressure equalization valve with his right hand.

The cabin had already lost most of its oxygen and pressure in the 15 seconds that had now passed since they separated from the orbital module. They continued their free fall through space toward the upper layers of atmosphere. Viktor, like Georgi and Vadim, experienced the results of rapid decompression in space. Thousands of small pockets of air and gas trapped inside the sinus cavities of his face had expanded, shattering much of the bone behind his cheeks and forehead. The oxygen in his blood was boiling away, escaping through his veins and arteries into adjoining tissues. Every cell in his body, now unrestrained by the pressurized atmosphere, was expanding. His arms and legs had ballooned to half again their normal size!

But they were still conscious and capable of action, if only for half a minute more. Viktor knew that recovery from rapid depressurization was possible if they could quickly restore cabin pressure. Decompression to a vacuum state was an unimaginable assault on the human body, but survival was quite possible. At least that was what he had been told during training. He knew they had to act fast, just as he knew that precious seconds had already been lost trying to figure out what was going on in the spacecraft.

Georgi and Vadim, seeing Viktor lunge toward the hatch, realized instantly that the pressure equalization valve was the source of trouble. They focused their attention on the atmospheric controls, yelling to each other and screaming at Viktor to move quickly as he began to close the valve above their heads. The atmospheric controls were largely automatic, and they knew all too well that it would take precious seconds - - maybe minutes - - to restore the atmospheric pressure in Soyuz.

Viktor turned the small handle connected to the pressure

equalization valve with his right hand as fast as physically possible, thinking only about grasping the small bar with his fingers, twisting as hard as he could, letting go, rotating his hand, and doing it all again each half second. He kept turning and turning. With each twist of the handle his chest collapsed and then expanded, sucking the last molecules of oxygen from the air that remained. Priceless time slid by with each rotation. His thoughts became even more frantic. "God, how many turns? Must be 20 or 30. Is it broken? No, it can't be. It's still turning and there is resistance. It's taking too long!" The seconds..10..20..and then 30..continued to pass by. "There isn't enough time! This can't be happening, it simply can't be happening." He looked down beneath his raised right arm and saw that both Georgi and Vadim were not moving and their arms were floating out away from their bodies. He looked back up to the hatch and realized that he could not continue to apply the necessary torque to the handle.

The movement of his wrist slowed and then stopped, but his chest heaved wildly. It was not going to work. He could no longer move. From the deepest reaches of his brain came a desperate cry, but there was only a neurological impulse. There was no air to inhale or exhale, no air to vibrate past his vocal cords, and no air in which to transmit the sound that could not be made.

Ground crews sat ready in helicopters, poised to take off to rendezvous with the spacecraft upon landing. As the Soyuz entered the uppermost layers of the atmosphere they could see a bright yellow scratch in the dark morning sky above. The heat shield was aglow from the friction of the air. The tension grew among the helicopter crews. They made their last preparations for flight and began to take off, first lifting up a few meters and then tilting forward and sliding above the flat ground and gaining altitude as they moved away. Their spotlights, focused on the earth, raced across the endless open field as they sped forward in the darkness. A few shouts of enthusiasm from dozens of people standing near the makeshift helicopter landing pads were heard above the roar of the turbine engines.

Without light the ground crews could not see the final stages of landing, but everyone at the scene knew the sequence well. At an altitude of nine kilometers a single drogue chute deployed automatically on Soyuz, followed by the single main chute at eight kilometers. With the main chute now deployed, the automatic sequencer on board Soyuz 11 jettisoned the heat shield, exposing the solid fuel landing rockets. The spacecraft and its billowing parachute were easy to see on the radar of the hovering helicopters. Soyuz 11 floated silently toward the ground from high altitude, far above the noise and commotion.

Soyuz continued her descent to the flat plane below and the helicopters raced toward the point of landing - - 1000 meters, 500 meters, 100 meters. She was right on target. The landing rockets fired two meters above the ground, making a bright flash in the darkness and kicking up a large cloud of dust. They cushioned the impact of the craft on the hard, dry soil. The capsule settled to the ground, rolled on its side, and the enormous parachute spilled its air into the darkness and collapsed onto the dirt.

The helicopters landed a short distance away, and the specially trained landing crews jumped out and ran toward the capsule. These men had prepared for this moment for years and were fully versed for all contingencies and emergencies. Upon first inspection the craft looked in excellent shape, despite the normal black scarring from the heat of reentry. Designated crews took their positions near and around the capsule while others radioed that the landing was successful. Camera and lighting crews jumped out of still more helicopters as they landed. Ground crewmen deployed reclining chairs next to the capsule for the three cosmonauts. They would not be expected or allowed to walk after such a long stay in zero gravity. The medical team was ready, as on all landings, and three local young women in elaborate native costumes stood by with large bouquets of camomiles to present to the men once they were outside. The entire area was now well illuminated by all of the aircraft at the site.

Two ground personnel released the locking mechanism and swung open the heavy hatch. There was no movement inside the capsule. They stood and stared in absolute horror, realizing that Viktor Patsayev, Vadim Volkov, and Georgi Dobrovolskiy were dead, sitting in their cramped seats. Investigators would later determine that one of the cosmonauts had managed to close the valve in the hatch only half-way before losing consciousness. The valve was intended to be used in just this type of emergency, yet it would have taken another minute of rapid turning of the handle for it to be completely closed and for the Soyuz to be air-tight once again. The precise conditions under which the control would be used had not been considered during design and construction.

Three and a half months later on October 11, 1971, Salyut ground controllers in Kalinin reoriented the Salyut 1 space station and fired her retro rockets. She slowly descended into the upper reaches of the atmosphere where the friction began to incinerate her metal skin. A long arching streak of light could be seen in the sky as she entered the atmosphere. The charred tons of metal that were once Salyut plunged into the Pacific Ocean after 2,800 orbits of earth.

REFERENCES AND NOTES

Brown, P.T. (1972). *Astronauts eye.* London: Obelisk Press.

Cause sought in Soyuz tragedy (1971). *Aviation Week & Space Technology*, July 5, 12-15.

Chaikin, A. (1985). The loneliness of the long-distance astronaut. *Discover*, February, 20-31.

Daniloff, N. (1972). *The Kremlin and the cosmos.* New York: Knopf.

Depressurization of Soyuz confirmed by Soviet commission (1971). *Aviation Week & Space Technology,* July.

El-Baz, F. (1977). *Astronaut observations from the Apollo-Soyuz mission.* Washington, D.C.: Smithsonian Institution Press.

Ezell, E. & Ezell, L. (1978). *The partnership: a history of the Apollo-Soyuz Test Project.* Washington, D.C.: NASA.

Furniss, T. (1983). *Manned space flight log.* London: James Publishing Company, Ltd.

Garelik, G. (1986). Our boys are dead. *Discover*, April, 59-64.

Johnson, N.L. (1980). *Handbook of Soviet manned space flight* (Science and technology series, volume 48, a supplement to *Advances in the astronautical sciences,* American Astronautical Society). San Diego, California: Univelt, Inc.

Lebedev, V. (1990). Diary of a cosmonaut: 211 days in space. New York: Bantam Books.

Orberg, J.E. (1981). *Red star in orbit.* New York, Random House.

Parker, J.F. & West, V.R. (Eds.) (1973). *Bioastronautics data book.* Washington, D.C.: NASA.

Raised Salyut orbit may signal new visit (1971). *Aviation Week & Space Technology,* July 12, 16-17.

Riabchikov, E. (1971). *Russians in space.* New York: Doubleday.

Ryumin, V. (1980). *175 days in space: a Russian cosmonaut's private diary.*

Salyut flight stresses biomedical studies (1971). *Aviation Week & Space Technology,* June 21, 17-19.

Salyut mission starts research on astrophysics (1971). *Aviation Week & Space Technology,* June 28, 22-23.

Salyut orbital altitude drops (1971). *Aviation Week & Space Technology,* August 2, 16.

Soviet control room shows tracking net (1971). *Aviation Week & Space Technology,* January 6, 15.

RENTAL CAR

Pilot Dan Bowman, United States Army Air Corps, snuffed-out his cigarette in the ashtray on the table and grabbed his flight gear as he ran out the front door of the quonset hut. Thirty-five Japanese planes had been picked up on radar only moments before. They were headed directly for the airfield.

Bowman estimated his distance from the row of parked AAF fighter planes to be about 300 yards. The air raid siren and impending attack caught him somewhat off guard. He usually tried to stay closer to the ready-room and runway, but today he was off talking with a friend at the supply depot. It was 11 o'clock in the morning, an odd time to come under fire.

The radar system gave the pilots a few minutes advance warning. It was imperative that they get the planes off the ground and into the air. The AAF fighter planes would be easy targets for the Japanese bombs and bullets if they were sitting on the ground when they arrived.

Bowman broke into a run headed straight for the small grouping of P-47 Thunderbolts which he could barely see through the stand of palm trees ahead. The ground was hard and white and flat, and he zigzagged between the tree trunks as he ran. The humid midday heat had already settled down on the small Pacific island. The shadows from the palm fronds overhead didn't seem to help at all. It was oppressively hot, and Bowman was building up a serious sweat as he reached the halfway point to the airfield. Only a hundred or so yards left to go.

Some pilots had already arrived at the planes nearest the ready-room where the crew usually hung out. Support teams were making quick exterior checks of equipment. The first engine fired up in the distance. Everyone always wanted to be the first into the air. No one wanted to be the last.

Bowman broke from the last few trees at the edge of the runway and into the open. More than half of the planes were already manned. There, 50 yards away, was an unattended P-47. He had flown this one many times before. Another pilot ten yards ahead and to the side ran to the plane and pulled the chocks from the wheels. Bowman stopped and looked around again. Another plane would have to be found. He was not in the mood to play musical chairs.

But the base had exactly the same number of pilots on hand as aircraft. There *had* to be one P-47 without a pilot. All he had to do was find it. *There* she was at the far end. It was the new craft that had been delivered only two days before. And no other pilots were running toward her!

Bowman's feet slid on the sprinkling of loose sand as he dashed under the wing and pulled out the wheel chocks. He kicked them aside and hurriedly pulled himself up and onto the top of the left wing. The sliding canopy was ajar about an inch. Bowman slid it back and dropped quickly into the seat of the sparkling new plane.

Something was not right. Something *really* was not right. Jesus! The whole cockpit was different! It *couldn't* be all that dissimilar from the earlier models. All he needed to do was catch his breath and sort things out. Yes, that was it. He buried his head down in the cockpit and scanned the instruments. "Let's see," he mumbled hurriedly. "There's the altimeter. No.....wait....ah...it's..the manifold pressure gage. What the... Oh, there's the altimeter. Fuel....fuel....where's the fuel? Damn it. This is insane."

Bowman heard the roar of the powerful engines around him and felt the prop wash of all of the AAC fighters moving off down the field to get into the air. His shoulders were hunched forward, his eyes glued to the unfamiliar control panel. He couldn't think. It was just too much to fathom. He cocked his head up and looked out through the open canopy. "Oh my God - - everybody's underway." The last few planes were starting their roll down the airstrip. Others were already climbing to altitude to meet the oncoming band of Japanese attackers.

"Start the damn thing. Just get it started," he screamed to no one but himself. "Ignition switch....ah....ah no! It can't be. Ah, here it is. Let's see.."

The first bomb fell from the sky and landed near the operations building not more than 100 yards to Bowman's left. He ducked reflexively and looked up out of the cockpit again. Nothing from the explosion hit the plane. Bowman was in one hell of a mess, nothing but a sitting duck with his wings clipped.

His frantic attempts to identify the controls finally paid off. The big and burley Pratt & Whitney engine sprang to life, but Bowman didn't pause to congratulate himself. Up ahead, at the other end of the airstrip, was an approaching plane. It looked like a Mitsubishi Zeke fighter-bomber. Under normal circumstances the Zeke would be no match for him and a toughas-nails P-47 Thunderbolt. But at this particular moment he was at an obvious disadvantage.

Bowman looked up at the approaching Zeke, back down at the control panel, and back up at the Zeke. It was hopeless. There was no way in the world that he was going to get this thing up in the air and survive the aerial attack.

The staccato bursts of cannon fire came as no surprise, but the multiple impacts from the Zeke's 20 mm cannon slamming into the runway were downright terrifying. He had to act *now*. Bowman grabbed the hand throttle. At least *it* was right where God intended it to be! He pushed it forward and stomped down on the left rudder pedal. The 2000 horsepower engine roared and the P-47 Thunderbolt swung her tail around counterclockwise. Bowman evened out the rudder pedals and scooted off to the side of the airstrip just as the Zeke thundered by, not more than 50 feet overhead, her guns still ablaze.

That was it! He had to keep the ship moving. There was no way that he could get her up in the air in conditions like this. But he sure as hell could run her on the ground!

The same Zeke would probably be back for another pass, this time from behind him, perpendicular to the airstrip. Bowman hit the left rudder pedal again and swung the 10,000-pound ship back around 180 degrees. She darted across the runway and slowed near the edge of the palm tree grove. The Zeke roared directly overhead above the tops of the palm trees, its pilot unable to locate his handicapped target.

For the next ten minutes Dan Bowman raced his plane up, down, across, and around the airstrip on that remote island in the Pacific. He was a tempting target, and the Japanese fighter-bombers made countless strafing attacks. But the American pilot was able to dodge all of the bombs and all of the bullets. Their ammunition low and their fuel tanks running dry, the enemy planes headed back out to sea and their own base.

Dan Bowman eventually learned to fly his new P-47 Thunderbolt and confess his story to aviation researchers after the end of World War II in August of 1945. But he never did figure out why someone would redesign a fighter plane's instrument panel in the middle of a war.

REFERENCES AND NOTES

Craven, W. F. and Cate, J. E. (1955). *The Army Air Forces in World War II: Volume Six, Men and planes.* Washington, D.C.: Office of Air Force History.

Fitts, P. M. and Jones, R. E. (1947). *Psychological aspects of instrument display: Analysis of 270 "pilot-error" experiences in reading and interpreting aircraft instruments.* U.S. Air Force Air Material Command, Wright-Patterson Air Force Base, Dayton, Ohio, Aero Medical Laboratory, Engineering Division.

Gurney, G. (1962). *The war in the air.* New York: Crown Publishers, Inc.

This story is based on a 1947 interview with an anonymous fighter pilot who served in the Pacific during World War II (see Fitts and Jones, 1947).

A MEMENTO OF YOUR SERVICE

It was at the conclusion of the small discharge ceremony on Pastrengo Rugiati's last day of service in the Italian Navy when the old captain approached and motioned for him to hold out his hand. Into it he placed the treasured object. It was the Captain's maneuver whistle and chain, worn about his neck for an entire career at sea. The Captain had been hard on him this past year, but Rugiati always sensed the old man's special interest in his progress on board ship. Presenting the memento was clearly important to the Captain, and Rugiati was equally taken by the gesture. For one, it symbolized the passing of the torch. For Rugiati, it was a heartfelt sign of his superior's trust in him, a trust that exists only when a man's life rests in the hands of another, a trust developed during a year at sea. The young officer listened carefully to his Captain's words.

"Here is a memento of your service," the old man said. "I had it with me when my ship went down." Rugiati knew that the captain had lost a ship soon after the end of the great war. "It was on the rocks off the Scilly Isles." Rugiati humbly placed the chain around his neck and bid his friend farewell. He would wear the memento for the remainder of his career at sea.[1]

Thirty years later at 6 o'clock on Saturday morning, March 18, 1967, Pastrengo Rugiati, Captain of the thirteenth largest merchant ship in the world, awoke to the ring of the telephone at his night stand. He was in his five-room suite directly below the bridge on the massive supertanker and had retired to bed just a little more than three hours before. The old Italian Navy maneuver whistle dangled from a chain around his neck as he reached for the telephone.

His answer was gravelly and half awake. "Yes?"

"It is 6 o'clock, Captain." The call, as expected, was from Silvano Bonfiglio on the bridge. Rugiati had left written instructions the night before that he be contacted when they picked up the Scilly Isles on radar, and he was to be called at 6:00 a.m. with a report if the islands had not been detected by that time.

Bonfiglio paused a moment to let Rugiati find his bearings after his deep sleep. Then he continued: "I have not seen the

---

[1]This story is told best in the book by Richard Petrow. Pastrengo Rugiati joined the merchant marines after his discharge from the Italian Navy around 1937. He re-entered the service at the outset of WWII and served first on one of the few Italian submarines to survive the Atlantic sea battles. He subsequently was assigned to a destroyer, the *Impavido*, which was seized by Germany for her own uses. Rugiati spent the remaining days of the war in a German concentration camp in Poland. He re-entered the merchant marines at war's end and assumed his first command in 1952 on the Liberty Ship *Italo Marsano*. During the next ten years he served on the *Buiba*, the *Golfo di Castellammare*, the *Arickaree*, the *Scherzo*, and the passenger liner *Homeric*. He became an officer for the Union Oil Company of California in 1963.

Scillies on radar."

Bonfiglio had obviously switched on the Raytheon 1400 radar unit. It had a maximum range of 40 miles. Rugiati asked his first officer when, based on their current position, they might detect the islands on the scope. Bonfiglio answered that it would be about 7 o'clock. Rugiati told him to call back when he made contact. He hung up the telephone and began to plan for the busy day ahead.

The supertanker had now passed the coast of France and was headed for the Scilly Isles 21 miles off Land's End, Cornwall, at the southwestern tip of Great Britain. At her current rate of 16 knots, the ship would pass the islands and the nearby rocky shoals in about two and a half hours. From there she was to continue up the west coast of England to the Welsh port of Milford Haven where her full load of 119,193 tons of crude oil would be unloaded. She had carried it all the way from Kuwait. Representative of the new breed of supertankers, she was much too large to pass through the Suez Canal. Her course had taken her down and around the Cape of Good Hope, up the west coast of Africa, and on to Britain. The autopilot had been engaged during most of the long voyage, and it had been set to steer a course of 18° since passing the Canary Islands four days before.[2]

[2]The ship, like many supertankers, operated within a strange web of companies and agreements, all designed to minimize taxes, fees, and expenses. She was well run and well maintained, however. She was built by Newport News Shipbuilding in Virginia in 1959 and later expanded at the Sasebo shipyard in Japan to twice her original tankering capacity. Her owner was the Barracuda Tanker Corporation of Bermuda, a subsidiary of the Union Oil Company of California. On

Rugiati's plan was to take her around to the west of the Scilly Isles and the nearby rocky shoals, then on up to the port in Wales. But there had been an eastward current running during the last hours of the 27-day trip, and the ship had veered about eight miles to the east of the planned course. The Sperry Automatic Steering System had kept the ship pointed 18° north-northeast, but they had drifted slowly to the right during the last few hundred miles. This was not that unusual considering the currents in that part of the Atlantic and the length of their 18° set.

The Captain was in a hurry to get to the unloading terminal at Milford Haven. Four days before on March 14, the company's agent in Britain had contacted Rugiati to tell him about the high tide on the night of the 18th. The supertanker had a deep draft, and she could barely make it into the shallow harbor entrance on the evening tide. It was going to be a tight squeeze even with the highest point of the tide at 11 p.m.

The supertanker, nearly 1,000 feet in length, was so heavily laden that she sagged in the middle. She would scrape the harbor bottom amidships even at the high tide at 11:00 p.m., something far too dangerous to risk. The solution, together with the high tide that evening, was to redistribute the crude oil toward her bow and stern so that her bottom was perfectly straight and level. This was done by pumping crude from the tanks amidships into the top few feet of usable space (called ullage) in the forward and aft tanks. Most of the sag had been eliminated by redistributing cargo during the previous week, but the work remained to be completed. The pumping would

---

her bow was the large, orange and blue Union 76 corporate logo, yet she sailed under the Liberian flag. Her all-Italian crew had been recruited by the Consulich Company, the Italian agents for Union Oil in Genoa. The tanker was under a one-time charter to British Petroleum for the voyage from the Persian Gulf to the tanker terminal in Wales.

take five hours, and Rugiati wanted the job to be done while the ship sat still outside the harbor in Wales so as not to slosh the crude oil onto the decks of his tidy supertanker.

All of this meant that they must arrive no later than 6 p.m. that day, and Captain Rugiati's estimate of their earliest possible time of arrival was 4:30 p.m. They would have to wait for six days for another tide of sufficient height if they failed to meet the deadline. The cost of such a delay, Rugiati knew, was substantial, and he was not interested in having six days of idle time marked on his record with the company. The crew would still have to be paid, the terminal would have to be reserved, and the great ship would sit with her valuable cargo.

The telephone in the Captain's quarters rang again at half past the hour. It was Bonfiglio, and the Scilly Isles had finally come into view on the radar. Rugiati listened on the telephone as Bonfiglio gave his report at 6:30 a.m.

"Captain, I have tracked the Scilly Isles on radar. We have moved over to the right of the course. I have headed the bow for the Scilly Isles."[3]

[3]Bonfiglio had detected the Scilly Isles on radar and observed that they lay ahead and to the left instead of ahead and to the right. He then took a bearing to determine the position of the ship and recognized that the vessel was some miles to the right of its intended course. Accordingly, before telephoning the Captain, he had called the helmsman to the wheel, taken the supertanker off of autopilot by moving the control lever from "automatic" to "manual," and commanded the helmsman to come to a course of 6°, nearly due north. His intention was to keep the ship on her planned track around the west of the Scillies. However, his navigation was slightly in error, and

Rugiati was angered. He had left specific instructions for the first officer to call him when the Scilly Isles came into view on the scope, not for him to change course. Rugiati had plotted his 18° course to take the ship to the left of the Scilly Isles, and he was incensed that Bonfiglio had altered course without his permission. Besides, Bonfiglio had not been especially clear about the reasons for his action.

Rugiati immediately questioned his first officer. "With our original heading of 18°, would we be free of the Scillies?"

Bonfiglio's reply was direct. "Yes."

"Then continue on course 18 degrees," ordered the Captain. "I intend to pass to the starboard of the Scilly Islands." They had obviously drifted somewhat off course due to a current, but, regardless, Rugiati did not like someone else changing the course of the ship. Sticking to the original 18° set and going to the right of thc isles was the preferable action.

"Pay attention," barked the Captain prior to hanging up. "In a few minutes I will be on the bridge."

Bonfiglio acknowledged and put the ship back on autopilot control with a heading of 18°.

A few miles to the right of the Scilly Isles lay the infamous Seven Stones, a collection of rocky reefs. The Seven Stones towered tens of feet into the air when the tide was low. Other times, as on this morning, the tide was relatively high, and the sea covered the menacing pinnacles. Appropriately enough, there was a large lightship anchored at the far side of the Seven Stones, and her 600,000 candlepower beam could be seen for 11 miles in every direction.

---

the ship was not quite as far west as he believed. The result was that his "correction," instead of steering them to the west of the Scillies, put the islands directly ahead.

Dressed for the day, Rugiati walked up the ladder and onto the bridge. It was a little after 7 a.m. The bridge of the supertanker was massive; there were only a dozen ship's bridges this large in the entire world. It ran the width of the vessel, but was relatively narrow front to back. Large windows spanned the sides and front, providing a complete view of the ocean and deck (most of which was well forward of the bridge). Banks of controls and displays lined the forward console below the windows. An open platform (a wing) sat off the left (port) and right (starboard) sides of the room for taking bearings and observing traffic.

Rugiati said good morning to Bonfiglio and went immediately to the radar scope to check the position of the vessel. As he expected, the Scilly Isles lay well ahead off the port bow and the ship was on a course of 18°. The autopilot would take them to the right of the Scilly Isles and her lighthouse, but to the left of the Seven Stones shoals and the lightship anchored nearby. It was a channel about six miles wide, large enough, he reasoned, for his tanker.

Rugiati had sailed around Cornwall in southwest England a number of years before while serving as first officer on the passenger liner *Homeric*. He had taken the route through the 21-mile wide channel between the Scillies and Land's End during each of the 16 passages through these waters, always staying clear of Seven Stones. He was not the master of his ship in those days, however, and his first-hand experience with the area was limited. But he reasoned that he could take the ship off of autopilot as they neared the channel between the Scillies and Seven Stones, set a course of 325° in order to navigate the channel, and then return to the 18° course.[4]

---

[4]It is agreed generally that only the ill-informed captain would ever try to navigate a big ship through the channel between the Scillies and

The Captain and his first officer spent much of the next hour discussing the problem with the sag in the ship, the cargo redistribution procedures, and the importance of getting to the terminal in Wales early enough to redistribute the cargo and enter the harbor on the tide. They calculated their position at 7:09 a.m. and then again at 7:45. The 7:45 position was based on a single bearing to one of the Scilly's lighthouses and a distance measurement from the radar scope.[5]

The supertanker plowed ahead, her course held firm by the autopilot. Each passing minute brought her 500 yards closer to the Scillies. Beyond the massive glass windows on the bridge, beyond the expansive bulk of the ship out front, lay the islands and the nearby rocks, slowly coming into view at the distant edge of the morning horizon.

Third officer Alfonso Coccio, 27 years of age, entered the bridge shortly before 8 a.m. He was there to relieve first officer Bonfiglio at the end of his four-hour watch. It was Coccio's first trip aboard the supertanker, and Rugiati was well aware of the young man's interest in his duties. True to form, Coccio had studied their intended course the night before and familiarized himself with the landmarks on which he would take bearings during his watch. Rugiati appreciated the fact that his young

---

Seven Stones. The standard navigation maps for this area off of Land's End all direct ships to steer clear of the Scillies and the Seven Stones and to steer a course well to the west or out into the deep and wide channel between the islands and Land's End. Rugiati had only one map of the area in the chartroom, and it did not provide the level of detail necessary to navigate under close quarters in these waters.

[5]This was a short-cut procedure not often used during navigation close to shore.

officer had prepared for his watch, and he sensed Coccio's disappointment when Bonfiglio briefed him on their position and the new plans, plans which no longer included sailing to the west of the Scillies. Bonfiglio took a fix on their position and plotted it on the map. The position he marked on the chart showed the supertanker to be five miles southeast of the nearest of the Scilly Isles. Bonfiglio then went below, leaving the bridge to Rugiati, Coccio, and the helmsman. The autopilot still pointed the ship on her course of 18°.

Rugiati removed a pack of cigarettes from his pocket and offered one to Coccio. A good smoke was a fitting way to begin a morning watch, and he enjoyed the young officer's company. True, Coccio was a little green, but he would learn the ropes quickly. And he was less likely than Bonfiglio to take issue with the actions of his Captain. Rugiati inhaled long and hard and leaned back on the bulkhead, blowing the smoke out toward the ceiling. The strong cigarette was a pleasant counterpoint to the forlorn hours at sea at the break of day.

There were footsteps coming up the ladder. They belonged to Biagio Scotto di Carlo, the helmsman, arriving for his 8 o'clock watch on the bridge. Scotto, like some of the other crew members, boarded the supertanker just prior to this voyage, but Rugiati had gotten to know him reasonably well during the past month. He seemed to be an experienced and able-bodied helmsman, and Rugiati had not been surprised to learn that Scotto had spent nearly all of his adult life at sea.

Captain Rugiati greeted Scotto when he entered the bridge and then politely instructed him to go back down to the deck below to get two ashtrays, one for him and one for Coccio. He delivered the ashtrays within a minute, and Rugiati sent him downstairs again on another errand. He returned a few minutes later and took his position at the helm, relieving the man on duty.[6]

The new watch settled into the morning routine. At 8:12, just as the ship passed by the first of the Scilly Isles, third officer Coccio took a bearing from the left wing of the bridge as St. Mary's Lighthouse passed exactly 90° to port. Rugiati watched Coccio as he re-entered the bridge and the adjacent chart room to plot the position. Rugiati followed to observe. The lighthouse was 4.5 miles abeam. Rugiati nodded his approval as Coccio marked the time (8:18) and an "X" for their position. It was a pleasure for Rugiati to watch his young officer learn to navigate a big ship. They were now sailing along next to the Scilly Isles and were 30 minutes distant from the Seven Stones. The sun was well above the horizon, and visibility was ten miles.

A few minutes later Rugiati noticed two small blips on the radar screen. Coccio obviously noticed them as well, and he walked to the scope for a closer look.

"Captain. There are two fishing vessels ahead."

"Yes, " replied Rugiati. "I have seen them already."

It was now time to change course and enter the six-mile wide channel between the Scillies and the Seven Stones, so Rugiati walked over to the helm where Scotto was standing. As usual, the ship was on autopilot. On the helm console, just to the right of the ship's wheel, was a lever with three positions. The first or near position (the one in which it was currently placed) was labeled "automatic." It was used to engage the autopilot and to automatically keep the ship on its current

[6]Neither Rugiati nor the helmsman briefed Scotto on their current position or the navigation plans.

heading based on the gyro compass. The second or middle position was the "manual" setting and was used when steering the ship by hand. The third position, the one all the way forward, was labeled "control," and it was used to disconnect the helm from the rudder. The lever would be placed in this "control" mode if the helm controls ever failed. This would enable a separate "control" lever on the left side of the helm console to steer the ship. For obvious reasons, it was seldom if ever used.

The large wheel at the helm, Rugiati knew, was not connected directly to the ship's rudder. Turning the wheel merely modified the electric signal to the console and the electronic devices it contained. The console, in turn, transmitted a signal to hydraulic actuators down in the bowels of the supertanker that pushed the massive rudder left or right. The lever on the console determined how, if at all, these inputs were transmitted to the rudder. The internal computer held the ship on its current course when the lever was set on "automatic" mode - - just as the cruise control on an automobile can be set to maintain a constant speed. When set on "manual" mode, however, the outputs to the rudder tracked the inputs made to the wheel.

A major change of course could be made only by pushing the control out of "automatic" and placing it in the "manual" position. However, a unique characteristic of the Sperry Steering System was that the pilot could change the course up to three degrees while the control was in "automatic" and the ship would hold to the new track automatically. So Rugiati, without changing the mode from automatic to manual, turned the wheel to the left and changed the heading from 18° to 15°. The big ship responded and slowly began to assume her new but only slightly modified course. A loud click could be heard from the gyro compass each time she passed through one degree of course change. It was Rugiati's feedback that they were turning. A small turn such as this brought about a few clicks; a big turn brought about a long span of clicks. The turn completed, Rugiati stood back again and monitored the progress of the ship.

He watched Coccio walk to the wing of the bridge and take another set of bearings minutes later at 8:25, then followed him to the chart room and looked over his shoulder as the younger man marked their position. It was time for another course correction, so Rugiati walked back to the helm and turned the wheel to the left to put the ship on an autopilot course of 12 degrees.[7]

At 8:30 the supertanker was still making her sweeping left turn to enter the six-mile wide channel between the Scillies on her left and the Seven Stones ahead and to the right. Two lighthouses on the Scilly Isles, as well as the lightship anchored near the Seven Stones, could be seen from the bridge. There was

[7]Everything, as far as he knew, was going according to plan. But their position was tenuous, at best. The first issue, of course, was the islands, shoals, and the channel ahead. It was not prudent to be taking a supertanker through these waters, but Rugiati had years of experience and no reason to think that he was about to get himself into trouble. Second, he had these fishing boats out in front. He was not interested in running through their nets (although they presented no danger to his massive ship), and he certainly did not want to run over the boats. Third, the eastward drift that had placed the supertanker off course in the first place was still running strong, moving them to the right at a rate of over one mile an hour. Fourth, he had to consider the "head reach" of his ship. A big ship, especially a supertanker full of oil, needs a lot of time and room to turn. Rugiati's ship would travel 1,550 feet forward (the "head reach") before it would complete a 20 degree turn and head off in the new direction. He had to consider this when he made his turns, and it was of obvious importance when maneuvering in close quarters.

much to which he needed to attend, but Rugiati was concerned primarily about the two fishing boats in his path. He had to maneuver around them, not an easy job considering the limited maneuverability of his ship.[8]

Suddenly, he spotted the line of orange fishnet buoys off the port side of the supertanker. There were others on the right side as well. Knowing that there was no alternative at this late point, Rugiati stood on the bridge as the supertanker sliced through the nets, her bow snapping the lines like thread.

Having made yet another modification to the course held by the autopilot, Rugiati stood by the helm as the ship passed through a 10° heading, continuing her sweeping left turn. But suddenly there were more fishing buoys to port. With the ship still on autopilot, he put her back on a 13° heading in order to avoid the nets. This brought her to the right side of the channel.

Third officer Coccio walked past to the port wing, took a fix on one of the Scillies lighthouses, walked back to the radar scope to obtain a distance, and then back to the chart room.[9] Rugiati followed him to the chart table and glanced over his shoulder. Coccio marked an "X" on his calculated position. A rush of anxiety, something he had never felt, hit him before he could verbalize his thoughts. How could it be? Coccio's fix on the chart had to be at least a mile off their actual course. It was physically impossible for them to have covered the marked distance during the last few minutes. He realized immediately that the inexperienced third officer should not have been

[8]Actually, these were two French "crabbers" working the shallow waters of the shoals. This alone should have alerted the supertanker crew of the danger ahead.

[9]Given the circumstances, it can be argued that the practice of obtaining a fix with a single point and the radar distance was entirely inappropriate.

entrusted with the navigation duties during the morning's passage. There was a good chance that they were nowhere near the position marked on the chart!

"Stop using the Scillies for bearings," he shouted at the junior officer. "Use the lightship." It was imperative that they fix their position based on two actual bearings, one of which had to be the Seven Stones Lightship. Rugiati looked out the wide windows to the lightship up ahead and then back to port toward the Scillies. He was not exactly sure where they were relative to the Seven Stones. All the while the autopilot held the ship to the last programmed course.

Coccio ran across the wide room and out the door to the starboard wing of the bridge. He was taking a bearing to the Seven Stones Lightship. Rugiati moved quickly to the radar scope and read the distance - - it was only 4.8 miles. Together they ran back to the chartroom and marked the bearing and their position. They were only 2.8 miles away from the South Stone, the southern most reef of the Seven Stones! It was now 8:40, and they had gotten themselves into serious trouble.

There was still time to steer the ship out of danger. Rugiati dashed to the helm, moved the rotary lever from "automatic" to "manual," and spun the wheel to the left. The bow of the supertanker turned slowly to due north (000°), and Rugiati reached down to move the mode control back to "automatic."

Coccio ran back inside from the starboard wing. He had taken another fix on the lightship. Rugiati watched as Coccio stopped at the radar set to obtain the distance. But then Coccio turned and ran back out to the wing and began to take another bearing. He had obviously forgotten the bearing in the midst of all the excitement.

Coccio hurried over to the radar scope again, then back to the chart table. Rugiati stood by as Coccio marked their position on the map. Coccio, the pencil still in his hand, looked up at

Rugiati. Their eyes met. Nothing needed to be said. They were less than three miles from the lightship. More importantly, a submerged reef lay less than one mile ahead directly off the bow!

Rugiati ran from the chart room around the corner to the bridge. Helmsman Scotto, who had been relegated to an observer since starting his watch, was standing out on the starboard wing. Rugiati screamed as he ran to the helm. "Come to the wheel! Come to the wheel!" The Captain reached down, shoved the steering system control out of the "automatic" mode, and turned the wheel counter clockwise with Scotto's assistance. "Hard to port," he shouted. "Go to 350." He decided that there was no reason to be conservative. "No, take her to 340. Take her to 320..." It would take minutes for the supertanker to complete the turn, but they had caught her just in time. The ship should soon be clear of the shoals.

Scotto had taken the wheel and was carrying out the order, so Rugiati turned and darted back into the chart room. He examined the map and saw that the new course would head them back out to the channel and well away from the Seven Stones. It had been a close call, but the quick course correction would put them back out into deep water. Never in his life, not even during the war, had he been through anything like this. It had been one thing after another for the last 20 minutes. This was one voyage that he would be glad to see come to an end.

Perhaps they should have just steered well to the east of the Scilly Isles, or perhaps to the west as he had originally intended. Regardless, they had made it through, and they would never again come so close to disaster!

Still, Rugiati felt uneasy. The course correction had been made well over a minute before, but things were not complete. Everything was fine, yet something seemed wrong. Yes, something was missing. A full minute after entering the chart room, more than a minute after issuing the last command to save the ship, he realized what it was. It was *too quiet*. Where was the clicking from the gyro compass? Each degree of the course change should have resulted in a loud click that could be heard on the bridge and in the chart room. But it was all quiet except for the normal rumble of the ship making her way over the ocean. And here, just when he thought they were out of danger, just as he finally had the situation under control, disaster might again be dead ahead. He ran once more through the doorway onto the bridge, only to come face to face with helmsman Scotto, who was headed his way.

"She's not turning, Captain," screamed Scotto. The supertanker had not moved from her course during the last minute. She was still headed straight for the rocks off the Scilly Isles. Scotto had been turning the wheel but nothing, absolutely nothing, had happened.[10]

Rugiati ran straight to the helm and spun the ship's wheel to the left. It was obvious that nothing was happening and that the ship was not responding. They had been maintaining their perfectly straight course all this time!

He had to move fast. It had to be one of the fuses. They might have blown one out when they turned the wheel rapidly minutes before. Rugiati quickly opened the small door covering the fuse panel. The three fuses were inside. Each had a test light. He bent down, reached inside, and pressed each of the three test buttons in turn. The ship drew 24 feet closer to the Seven Stones with each passing second. He pressed the first

[10] Scotto had shouted to Rugiati and Coccio for them to come and help, but they could not hear him back in the chart room.

little button, then the second, then the third. A small red light came on with each action, signaling that the fuses were fine.[11]

What could it be? Why wasn't the ship turning? More than a minute and a half had now passed since he had first instructed helmsman Scotto to set the new course away from the Seven Stones, yet the supertanker continued her mindless, lumbering track toward the submerged rocks.

The oil pumps! That was it. It had to be the oil pumps! Rugiati knew that the oil pumps had broken down once before when the ship was new. The hydraulic system that powered the rudder could not work without the power provided by the pumps. The electric signal from the helm controlled the hydraulic system, which in turn powered the rudder. The electrical system at the helm was probably just fine.

Rugiati leaped towards the telephone near the helm, placing the receiver to his ear with one hand and dialing the engine room with the other. He waited for the connection and then for the ring. The view out the big windows toward the bow was unchanged. The supertanker kept rolling toward the submerged rocks somewhere ahead. Would they have enough time to start the auxiliary pump?

Someone answered the phone. It was the steward. What in the world was *he* doing down in the engine room?

"Ah Captain," said the steward politely. "Are you ready for breakfast?"

He had reached the officers' dining room by mistake! In his panic he must have become confused over the numbers and dialed a 14 instead of a 6. He slammed the receiver down into its cradle and cursed loudly.

It was madness! Of all the times in the world for the steering system to fail! They had to be close to the shoals by

[11] In fact, the steering system had blown a fuse once before, and Rugiati was somewhat predisposed to this conclusion.

now. Yet he stood helpless as the ship and her nearly 120,000 tons of crude oil kept barreling along toward the rocks off the Scilly Isles.[12]

Rugiati stood back a few feet from the helm to collect his thoughts and chanced to glance down at the steering control console. Of course! "Porco Dio," he shouted. The steering control lever for the autopilot, the one that had the three positions, was not in the "manual" position. It was set to the "control" position, the mode that they never used, a mode that disconnected the rudder from the wheel at the helm! He must have placed it there by mistake minutes before when he took the ship off autopilot.

He lunged toward the panel and set the steering control to "manual," turning the large helm wheel counterclockwise as quickly as possible with Scotto's help. Slowly, the bow of the *Torrey Canyon* began to swing to the left. The click from the gyro compass could be heard loud and clear as she passed through the first few degrees of the course change. She had turned left just 10 degrees to a heading of 350° when the first section of her hull struck bottom. The massive, ponderous ship ground over the submerged, razor-sharp ridges of Pollard Rock. The stone pinnacles tore lengthwise through the soft underbelly of the leviathan hulk, inflicting a gaping and lethal wound. She came to a painful and lumbering halt, and a thick river of black Kuwaiti crude began its unhindered flow into the cold, blue sea. In a few days the beaches of England and France would be choked with 31 million gallons of oil. It was a rude introduction to the age of colossal environmental disasters, and this was a disaster the likes of which the world had never seen.

Rugiati placed his left hand across his forehead, then ran his

[12] Rugiati, under the stress of the moment, never thought of slowing the ship or stopping the engines.

fingers back through his coarse hair. His right hand moved to his chest. Under his palm, beneath the sweater and shirt, lay the old chain and maneuver whistle given to him by his captain 30 years before.

REFERENCES AND NOTES

Cahill, R. A. (1990). *Disasters at sea*. Kings Point, New York: The American Merchant Marine Museum Foundation.

Cahill, R. A. (1985). *Strandings and their causes*. London: Fairplay Publications Ltd.

Cowan, E. (1968). *Oil and water: the Torrey Canyon disaster*. New York: J. B. Lippincott Company.

Gill, C., Booker, F., and Soper, T. (1967). *The wreck of the Torrey Canyon*. Newton Abbot: David & Charles Limited.

Marriott, J. (1987). *Disaster at sea*. New York: Hippocrene Books Inc.

Petrow, R. (1968). *In the wake of the Torrey Canyon.* New York: David McKay Company, Inc.

Pheasant, S. (1988). The Zeebrugge-Harrisburg syndrome. *New Scientist, January 21*, 55-58.

CHUTES AND LADDERS

David Goldman, accident investigator and expert witness, raised his hand and pointed out to his driver the approaching fence on the left. The car slowed and turned off the highway, passing gingerly through the open gate and on to the dusty and bumpy farm road. Rich Higgens, the young attorney working on the case, was visibly perturbed about driving his new BMW over the unpaved road. It was the kind of thing you needed to think about before going on a site visit like this, perhaps the kind of thing Rich would eventually learn, thought David.

The dirt billowed around the tires and settled to the glossy black exterior of the car. Rich leaned forward to look out the windshield and the dust on the hood.

"So much for the wash and wax."

The road ran straight back for a quarter of a mile and then cut squarely to the right, running parallel to another pole and wire fence. David checked his scribbled directions one last time. They were looking for the gate at the end. There, directly ahead, were the big overhead power lines, the location of the accident. Rich slowed the car and angled left through the opening, then stopped just inside the gate and turned off the ignition.

It was quiet outside, strangely quiet. Up above were the half-dozen live electric transmission wires that ran between the poles. They looked innocent enough way up there, but the lines were one of the main reasons for their trip out to the farm that day.

David enjoyed getting away from the office, especially when he could do a little work out in the open country. He knew that, on this particular job, the answers to his questions would most likely be found at the site of the accident; and he was much like a detective who speeds to the scene of a crime to find his clues. One of his biggest problems had been convincing the client of the need to spend time studying the site and interviewing the people involved. As on so many cases, the attorneys said that *they* would take care of collecting the facts; David's job was to analyze. They seldom understood the types of things for which he was looking or the bias in their own views. You had to put yourself in the victim's shoes, understand what it was that they were doing, see what they had seen. This was something that could not be done by listening to second- and third-hand accounts from people who had a financial interest in the outcome.

The attorneys had racked-up hundreds of hours concocting reasons "why" and "how" the accident had occurred. Yet none of them, amazingly enough, had taken the time to come out here and spend the day examining the setting and talking with the people who actually did the work. A man had died in this accident, and it was especially important that David understand exactly what had happened and why.

Rich walked around from the driver's side. His black-tasseled penny loafers and the cuffs of his suit pants were already sprinkled with dust.

"Well, I think this is the place," announced Rich. "John said he would watch for us and come over when we pulled up." John Weiss was the foreman on this section of the property. He was probably out in one of the nearby fields and would see them as they drove across the dirt road.

Sure enough, a white, late-model pickup was casually making its way toward them from a distant field to the west. David leaned back against the side of Rich's car and looked around the field. The soil and vegetation showed the tell-tale signs of recent activity - - vehicles and harvesting equipment had been through. The sky was gray and overcast, but it had not rained for much of the summer and the ground was dry. The hops growing in the field had probably been harvested within the last week.

Off in an adjacent field, on the other side of a long and narrow line of trees, he could see a group of men working. Like most of the farmworkers in this part of the state, these were probably Hispanic men. They appeared to be packing up long aluminum irrigation pipes lying in the field, but it was a little difficult to tell from this distance.

The local dry spell of the past two growing seasons had compelled farmers to undertake regular irrigation of their hops, and the only way to effectively water under such conditions,

David knew, was with sprinklers. Accordingly, the long, 38 ft. aluminum pipes had to be laid out, the field watered, and the lines disconnected and stacked on the long, lanky trailers. The cost of moving the sprinklers from field to field was less than purchasing multiple systems, so crews of men were kept busy disassembling, moving, and reassembling the pipes.

And that, of course, was the second reason why he and Rich were there. As had happened a number of times in this part of the state this summer, a farmworker, Arturo Salvador, had been electrocuted on this very spot a few months back. The source of power, as in all of the other electrocutions, had obviously been the power lines above. And there was no mystery about how the current had traveled from the power lines to Arturo Salvador: through one of the 38-foot aluminum irrigation pipes. The important questions in David's mind were the "how" and "why" of the tragedy. First, how did the pipe that Arturo Salvador held come in contact with the power line? And second, why did he lift the pipe so high in the air near the power line in the first place? It would be for the attorneys, a judge, and a jury to decide who was ultimately responsible for the accident, but it was his job to explain how and why it had happened. This was the only way to begin the process of addressing the problem and reducing the chances of future accidents.

The attorneys for the plaintiffs and the attorneys for the defense were locked in endless and often pointless battles over who was responsible for the accidents and who, if anyone, should bear the financial consequences. Some said it was due to careless farmworkers; some claimed that it was the power company's fault for putting the lines there in the first place. Some even maintained that the pipe manufacturer was responsible. And of course there were those who blamed the farmers. David's concern, however, was not blame or reparation. His interest was with Arturo Salvador's job and tasks on that fateful day and the circumstances that led to the accident. This was not an act of suicide, and there had to be a unique set of circumstances leading up to his death. All of the other arguments and discussions were of little concern to him, and he felt that most of the work to date had been pointless. The only important thing to him was determining why and how this accident and the others like it had occurred and what could be done to stop it from happening again. David believed strongly that there were simple answers to his questions as well as simple solutions to the problem.

There were so many questions that had not been answered - - or even asked - - during the "investigations" into these mishaps. Perhaps it had been the preoccupation with the monetary settlements or maybe the remoteness of the location. It was also possible, as some alleged, that the accidents had not been thoroughly investigated because they all involved migrant laborers. Whatever the case, he was determined to understand why these mishaps had occurred. It did not matter if he was working for the families of the victims, the power company, the farm owners, or even the manufacturer of the pipes. His job was to determine the circumstances and human actions that were leading to these electrocutions - - and, specifically, the electrocution of Arturo Salvador.

One thing David knew for certain was that the combination of the low power lines above the farm land and the long conductive irrigation pipes was disastrous. Like oil and water, some things were simply not meant to go together. As long as the power lines transversed the farms and as long as the pipes were long and conductive, there was the potential for fatal accidents. This case was not at all unlike others on which he had worked involving aluminum masts on small pleasure boats. With frightening regularity, unsuspecting sailors were being electrocuted in and around waterways when their tall, conductive masts contacted live overhead lines. The accidents provided lots of work for the attorneys and investigators, and the resulting education efforts with sailors apparently had some positive effects. But the real issue, David knew, was that these two conditions should simply not coexist, and, as long as they did, no warning sign, label, or so-called education program was going to eliminate these accidents. He suspected that the same was true of the electrocutions on the farms in this part of the state.

John Weiss, the foreman, drove his pickup through the gate and stopped next to their car. Rich and David walked over to meet him. John was courteous, but he obviously had other things on his mind and was visibly unenthusiastic about meeting them out here in the field. He wasted no time making his opinion clear - - the accident was due only to Arturo Salvador's own stupidity. There was not, in his opinion, much of anything for the attorney and the so-called accident expert to see out in the field, and certainly nothing to learn by driving out here. Arturo Salvador had lifted the pipe and touched the power line. It was as simple as that. He would show them around and answer any questions that they might have, but they were wasting their time.

There was little doubt in David's mind that the foreman was really there to see that he and Rich stayed out of trouble. John Weiss wasn't going to be of any help, but they had to get started somehow. Perhaps if he asked the foreman for his opinion he would leave him alone long enough for him to get some real work done. There were times when it paid to flaunt your ignorance.

"Well, can we at least hear what you know about the accident? This is the first time Rich and I have been out here, and we would like to get a feeling for what happened," said David.

"Whatever suits you."

David walked over to the long irrigation pipe lying on the ground near the truck. It was the very pipe that Arturo Salvador was holding when he was electrocuted. His first question was the most obvious one. "Do you know why he had the pipe up in the air?"

"Well, my guess is that he was moving it away from the gate so he could drive his truck through."

David had heard this theory before from one of the attorneys. True, this might have been a reason for moving the pipe, but it had little, if anything, to do with *why* he might have had the pipe up in the air.

David couldn't let this one go. "Uh huh. But, why would the pipe be up so high in the air if all he was doing was moving it. Can't you just pick up one end of this and pivot it around?"

"Well, hell, I don't know," snapped the foreman. "Who knows why he did it."

David continued. "Did he have any equipment out here? A lift or a crane or something? How could he have raised it that high into the air?"

"Well, your guess is as good as mine. I don't have all of the answers," said the foremen.

This guy didn't have *any* answers. And Rich didn't want to get his clothes dirty. David Goldman realized that he was going to have to figure things out by himself.

Rich and the foreman began their own conversation, so David walked along the length of the five-inch diameter pipe thinking about the accident and how Arturo Salvador might have lifted it up into the air. The foreman's argument that Arturo Salvador moved the pipe to drive out the gate was plausible, but it didn't make sense in view of the other evidence. There had been a number of these electrocutions that summer, and they all were certainly not caused by pipes blocking fences and roads. Plus, there would have been no reason to stand the pipe up into the air. There had to be another explanation, but he could not think of one. He would set that question aside for the time being and address the question of "how." More specifically, how did he get it into the air?

David reached the end of the pipe, slid the note pad into the back pocket of his jeans, stooped down, and lifted the pipe up to waist level. It was astoundingly light, much lighter than those aluminum masts he had tried raising while working on the boating electrocution case. Why not give it a try? He lifted the end up over his head and simultaneously stepped underneath it, arms straight up and elbows locked, the weight of the pipe resting on his palms. The other end was parked solidly down in the soft dirt 38 feet away. The pipe had quite a bit of flex. David then walked forward with the end of the pipe held over his head. With each step he moved his hands forward down the pipe. It got heavier as he continued, especially once he reached the halfway point, but in no time at all the back end of the long aluminum irrigation pipe was towering high into the air.

It was as sudden and as unexpected as it could have been. "Stop!" It was Rich, screaming at the top of his lungs. "Stop! Don't move!"

He looked up toward the sky, his shoulders and arms frozen in place over his head. Sure enough, the end of the irrigation pipe was about a yard away from one of the power lines overhead. He had forgotten completely about it, just as Arturo Salvador had forgotten about it the month before. The lines were well above and out of his normal line of sight, and he would not have thought of looking up, especially out here in a big open field. And this was perhaps the only field on the entire farm with a power line - - yet another reason to conclude that Arturo was not thinking about the danger overhead. David backed up a few feet away from the wire and let the pipe fall back down beside him. It reverberated loudly as it smacked and bounced up off the ground.

David walked back over to his two spectators. The foreman said nothing, but Rich could not resist the temptation.

"That would have been great, Dave. I can just imagine the headline: 'Safety expert electrocuted while reenacting accident'!"

It *had* been a close call, but David Goldman had now answered two of his questions. First, Arturo probably "stepped" the pipe up to a vertical position, just as he had done moments before. The pipe was not especially heavy, and raising it was something most any adult could do. He also understood *how* he might have done this without considering or seeing the overhead wires. Plus, Arturo was probably looking down at the end of the pipe resting on the ground, making sure that it was not going to pop up in front of him after passing the halfway point while he "stepped" the pole.

But the most puzzling question of all still remained unanswered. *Why* did Arturo have the pipe pointed up into the air in the first place? You could move one of these by lifting it up right in the middle. You could probably even balance and carry it on your shoulder by yourself. And there was no question that a person could move it a few feet quite easily by lifting only one end and pivoting it on the other. So why would he have wanted to stand it up in the air as David had just done? It didn't make sense.

David turned to the foreman. "Do you mind if I walk over there and talk to some of those guys?" referring to the crew in the adjacent field. They were disassembling the irrigation pipes and loading them on the trailers. David had always believed that the best answers were found at the source.

"By all means, help yourself," responded the foreman.

"OK. Then, I'll be back in a few minutes." He walked away and left Rich and the foreman standing next to the truck.

He was about 30 feet away when the foreman shouted out, "Hope you speak Spanish!"

David walked along the headland, next to the fence, until he reached the corner of the large adjacent field where there were about a dozen men working. They were disconnecting the long irrigation pipes and carrying them over to the trailers where they were stacked. He approached one of the long trailers where two men adjusted the pipes on top. It was all a little awkward. These men were, no doubt, friends with Arturo, and they might be suspicious and defensive. Should he try to introduce himself in his mumbling Spanish, or speak English? He just didn't want to be rude. His concern, as it turned out, was unjustified.

One of the two men near the trailer wiped the dirt off his right hand on his pant leg and extended it to David as he approached.

"Hi. How do you do?" said the farmworker in flawless English as they shook hands. "I assume you are here to investigate Art's accident?"

"Yes I am. May I ask you a few questions?"

"Sure. Whatever we can do to help. But you know that no one else was around when it happened."

"Yes, that's what I've heard." He paused for a moment and then continued. "There is really only one question that I have. No one seems to know why he was holding the pipe up in the air like that. Do you have any ideas?"

The farmworker smiled modestly, then raised his hand to eye level and wiggled his pointer finger to say "follow me." He turned and walked over to the end of one of the pipes lying on the ground about 50 feet away. He hesitated and turned to look back and then around to a few of his coworkers who had stopped what they were doing to watch the show. David thought he saw him wink, but he wasn't sure. Then he stooped down and grabbed the end of the pipe with both of his hands.

This was all *very* odd. These men were grinning. They obviously knew something he did not.

The farmworker lifted the end of the pipe to the level of his waist and on up over his head. Then, just as David had done minutes before in the adjacent field, he "stepped" the end of the pipe up into the air. Seconds later it was vertical, with the bottom end resting firmly on the ground. He looked around to the other men, then bent his knees and hugged the pipe firmly between his arms and chest. His knees straightened up and the bottom of the pipe raised about a foot off the ground, the 38 feet of aluminum pointed straight up into the sky. He twisted his torso and shoulders quickly left and right a few times, and the pipe began to wobble and wiggle.

Jostled loose from its comfortable haven deep inside, a plump beige and white rabbit plopped down out the bottom of the pipe, rear-end first, its hind legs wedged up under its chest. It landed on the dusty ground and twisted, trying to get its legs up underneath, but not before the other large ball of fur dropped down onto its head. The two rabbits rolled to their feet, hesitated for a split second, and scampered across the field in terror.

"Great entertainment!" said the field hand, as his friends laughed and turned back to their work.

REFERENCES AND NOTES

The names of the characters in this story are fictitious. Any similarity to their actual names is purely coincidental.

BUSINESS IN BHOPAL

Chairman Warren M. Anderson, Union Carbide Corporation, ducked his head as he emerged through the open doorway of the jet plane and stepped purposefully down the slippery metal stairs, one hand clutching the rail. It had been a very long day, and he was determined not to slip in front of the spectators and TV cameras. The last step down was the one onto the solid and seemingly reassuring gray asphalt of the airport tarmac. A slight breeze drifted by and his attention shifted involuntarily to the horrid stench in the air. It was a putrid mixture of jet fumes, industrial effluent, animal waste, and humanity; the smell of a living nightmare. Yet, there he was, right in the middle of it all.

The flights had taken him from the solace of wintertime Connecticut to this dirty and troubled city in central India, halfway around the globe. Images were more important now than ever before, but somehow the scene was not unfolding as planned. He flew here to demonstrate Carbide's concern, but it was not coming across well. There stood the wealthy American executive with his entourage in tow, intent on conveying his empathy to the residents of this sprawling third-world city, most of whom lived in abject poverty. His tailored clothes, gray hair, and heavy black glasses were just the finishing touches to the picture. He could not have looked more out of place, and he was not fully prepared to face what were to be perhaps the worst moments of his 63 years of life.

Everywhere, on the other side of the wire fence surrounding the field and through the windows of the buildings, were the small dark faces of people, hundreds and hundreds of people. They all looked alike, and they were all staring straight at him. There were no shouts or signs, but the indignation and infernal hatred of every person in the crowd was more than obvious.

His two foreign associates from the firm's subsidiary, Union Carbide of India Limited, stepped down off the stairs behind and joined him on the ground. Warren Anderson took a slow silent breath, and the three Union Carbide executives began to walk toward the crowd.

They had taken just a few paces when a group of five men emerged from the mob about 100 feet away and began walking toward them. Anderson's eyes turned quickly. This was obviously not a welcoming party of local politicians eager for their business. But something about their appearance suggested that they were not going to hurt them. They were reasonably well dressed, perhaps government officials. They continued walking toward the approaching men.

Their purpose became more clear as the group neared.

These people had their own point to make, and they were going to play it out in front of the media and the crowd. There were two agendas on the table - - his and theirs. But this was their turf, and it gave them an all too obvious advantage.

"Mr. Anderson?"

The group of five men stopped suddenly. The one who spoke was in front. Warren Anderson took two more steps, then stopped as well, face to face.

The accent was characteristically Indian, slightly high in pitch and staccato in its delivery. The man's voice was harsh. There were not going to be any pleasantries.

"Yes."

"We are officers of the Central Bureau of Investigations. Is this all of your party?"

Hearing who and what they were brought little surprise, but the abruptness of the question was disconcerting. This sounded a little more serious than anticipated.

"Yes."

There was a prolonged silence and everyone present absorbed the essence of what was about to be said, including, for the first time, Warren Anderson.

"Mr. Anderson, you and your associates are *under arrest* for conspiracy, criminal negligence, and criminal corporate liability as called for under Indian law. The three of you shall come with us."

Unbelievably, these people were really going to go through with this, and there was no choice but to go along. They were going to hold him personally responsible for what had happened at the pesticide plant! They were laying the blame on the parent company and, specifically, him!

Warren Anderson knew that the charges were part of an orchestrated media event, but no level of rationalization could quell the shock of being arrested. He knew also that the maximum penalty for the crime for which he had just been charged was *death*.

A few days before, around 10:45 p.m. on the evening of December 2, 1984, Suman Dey, an operator at the Union Carbide pesticide plant in Bhopal, assumed his position in the control room for the start of the graveyard shift. As usual, Dey thought it would be a long and generally wearisome night, and there was no reason to suspect the occurrence of anything out of the ordinary. Like so many nights before, he would spend most of the long hours strolling up and down the panels, reading gauges, and entering data in the log book. But, he had to admit, some of the time would be passed in conversation with the other operators. It seemed that every night for the past month they had talked about the deteriorating conditions at the plant.

They weren't accomplishing anything by talking about it, he knew; but there was no denying that things had never been worse - - and there was some comfort in sharing one's problems with others. The company was continuing to cut their budgets, and it was taking a visible toll on operations. It had also begun to make his own job more difficult. Much of the control room instrumentation that told him what was going on out in the plant had gone unrepaired for months, and reports of leaking valves and pipes outside were becoming a nightly occurrence. The most upsetting part of it all had to be the layoffs. Fortunately for Dey and the other operators inside the control room, the maintenance people had taken the brunt of the personnel cutbacks. True, the demand for *Sevin*, the pesticide they manufactured, had declined over the years, and the plant was operating at a loss, but the company could run the place better. Like so many of his trained colleagues, he would be gone by now if not for the incredibly poor market for his skills and knowledge. He should feel fortunate, to say the least, especially compared to the two hundred thousand people living in the shanty towns that had grown up around the plant in the last dozen years.

Dey began his nightly routine by walking to the high table in the middle of the room and opening the log book to review the activities of the previous shift. It summarized most of the day's major events, but the entries were often quite sketchy due to the operators' reluctance to write in English, as management required, instead of their native Hindi. A few hours before, according to the last scribbled entry in the log, the maintenance crew flushed the lines of the methyl isocyanate (MIC) unit with water. MIC, the main ingredient in the manufacture of *Sevin,* was stored in three large tanks outside. Flushing the lines was an operation that could be done only when they were not producing the pesticide, because the MIC was highly reactive with water. The long pipes connecting these tanks to other areas of the plant filled with residue over time and required flushing periodically with high-pressure blasts. The field operators, he hoped, had the wherewithal to show the new maintenance supervisor how the procedure was performed. Many maintenance jobs were being done poorly or not at all, now that the crews had been cut and the inexperienced people from the company's battery-manufacturing division had been put in charge.

A few minutes later at 11:00 p.m. it was time for Dey to walk back by the MIC panels and make his periodic check on the systems. He scanned the assortment of displays on the panels, including the pressure gauges for the three large underground tanks of MIC outside. The scale on the displays ranged from 0 to 55 psig, showing him the pounds of pressure per square inch above the normal atmospheric pressure. They had not manufactured any pesticide for many weeks, but it was still necessary to maintain a small amount of pressure inside the tanks in order to ensure that nothing, especially water, entered them. Tank 610 showed a pressure of 10 psig, just about in the middle of the acceptable range of 2 to 25 psig.

Unknown to Dey, the pressure inside the tank was 2 psig only 40 minutes before at 10:20. But the buildup was not apparent because no historical trace of the pressure was shown within the control room, and the operator on the previous shift had not entered this into the log.

As on every night at the plant, a small group of workers tended the maze of pipes, valves, pumps, and tanks outside the control room where Dey monitored the MIC systems. They had been quite busy outside for the past half hour, which Dey learned only when one of the newer field operators entered the control room around 11:30 p.m. The field operator was a short, middle-aged man, and he seemed like a well-meaning fellow, although he had not been on the job long enough for Dey to get to know him very well.

The operator did not appear to be too alarmed, just trying his best to make a clear report to the control room. Dey listened attentively as the field operator gave his report.

"It looks like we have an MIC leak outside. We weren't able to locate it at first, but then we heard it blowing near the scrubber."

Dey knew how MIC leaks were detected outside - - the

operators felt their eyes and then their throats and chests burn when there was a problem. Unlike their sister MIC plant in the United States, the Bhopal facility relied on the operators for leak detection. Throats, lungs, eyes, and ears were the gas sensors at this plant. There were no automated environmental monitoring systems outside, no gas sensors, no automatic alarms. They had to locate the source of escaping gas by seeing the plume or hearing a high-pressure screech. It was a difficult thing to do in the dark.

The field operator continued talking, and Dey listened carefully. "Well, when we saw the plume near the scrubber we went to the relief valve on the downstream side. Dirty water and MIC were flowing out the valve at a pretty good rate. The process safety valve has also been removed."

Dey, for the first time that evening, was moderately alarmed. He was also annoyed. He shook his head back and forth in mild disdain and looked upward momentarily, not at anything in particular. It was undoubtedly related to the maintenance crew on the previous shift and their washing of the lines. He thanked the field operator and told him to report back if anything else was found.

The MIC leak was relatively minor at this point in time and in and of itself was not of great concern. Small leaks, after all, were really quite common. What did disturb him, however, was the report that MIC and water leaked out of the relief valve. Under no circumstance should these two substances ever come into contact with each other. A mixture of water and MIC, he knew, could be very volatile.

Dey returned to the control room after a tea break about 45 minutes later. He did not talk about it during his break, but the persistent plume of MIC vapor outside and the discovery of water and MIC together in the connecting lines was bothersome. So, he walked directly past the 70-odd panels in the room over to the console containing the displays for the three MIC tanks. The pressure within Tanks 611 and 619 were just fine, but the pressure in Tank 610 was reading 25 to 30 psig, just beyond the edge of the acceptable range. When he had last looked at it, he recalled, it was 10 psig. Dey walked hurriedly across the room to talk with a colleague and get a second opinion. They both returned to the panel hardly a minute later and looked at the pressure gauge for tank 610. The needle was now "pegged" at the extreme end of the scale! Dey, as well as his colleague, knew this meant that the tank pressure had risen quickly to at least 55 psig - - and could be much higher, how much more they had no way of knowing. He headed directly toward the exit of the control room knowing that they might have a very serious problem on their hands.

Dey pushed open the heavy steel door, and the slightly cooler air from outside drew past him and into the building. Hesitating for a moment, he stopped and scanned the miles of pipes and equipment rising before him in the glow of the incandescent lights. There were no visible signs of trouble, but the burning in his eyes told him that they had a leak somewhere. He was concerned for his safety for the first time since beginning work at the plant.

The heavy steel door slammed shut with a loud bang, as if to rudely prompt him into action. Dey started running in the direction of the MIC tanks; each footstep on the ground made a distinctive crunching sound, the only sound he noticed until he covered a few dozen yards and approached the MIC storage area where a venting safety valve was screeching. Slowing to a fast walk, he took a moment to catch his breath and survey the MIC storage area.

For safety reasons and to help maintain a low temperature on hot summer days, each of the double-walled stainless steel 15,000 gallon tanks was buried underneath a large rectangular slab of concrete. Tank 610 was under the slab at the far end. He broke into a run again, stopping alongside the concrete slab and pipes that extended from the tank underground. The relief safety valve was the one that had popped. Its graphite rupture disk must have given way, as it was designed to do when the pressure got too high. His first concern was knowing the temperature in the tank, something that was not displayed to him in the control room. The storage system was designed to keep the MIC relatively cool, so the maximum value on the round temperature gauge mounted to the pipes was just 77° F. Incredibly, the needle on the gauge was "pegged" at the 77° F mark, meaning that the temperature was at least that high. He looked next at the pressure gauge attached to the line with the relief valve that was venting the gas above his head. It read 55 psig, again, the maximum value on the scale.

Dey placed one foot up on the concrete slab and stepped up on to it. Oddly, he lost his balance and grabbed a nearby rail to steady himself. Once again he sensed that he had slipped and then realized that the huge concrete slab was shaking! He looked down and saw that the concrete slab on top of the tank - - the slab on which he was standing - - was cracking apart beneath his feet. It meant just one thing: an unprecedented, runaway chemical reaction was underway inside the tank. It contained over 40,000 pounds of deadly methyl isocyanate (MIC)!

There was no way of knowing it at the time due to the inadequacy of his instrumentation, but the temperature in the tank was now well above 392° F. The maintenance crew that washed out the lines earlier that evening had failed, for a variety of reasons, to isolate tank 610 from the pipes they were cleaning with high-pressure water. Hundreds of pounds of water entered the tank and an enormous, volatile chemical reaction had been brewing ever since.

He jumped off the concrete slab onto the gravel and ran over to the vent gas scrubber a number of yards away where the MIC supervisor, Shakil Qureshi, and field workers were trying to figure out how to stop the MIC gas from escaping out the pressure relief valves. Dey shouted from some distance away as he ran toward them.

After a quick discussion, the small group of men rushed over to the MIC storage tanks. The concentration of MIC in the air had been increasing steadily, and Dey's lungs were starting to burn. Shakil Qureshi, relatively new to his job and generally unfamiliar with the MIC facility, surveyed the situation with Dey and then instructed the operators to isolate the leak by closing various valves in the network of pipes. There were no remote actuators in the control room - - all of the valving changes had to be made by the operators outside at the pipes, even when there was a dangerous leak such as this.

Their attempts were futile. The pressure within the pipes was far too great, and the gas blew past all of the valves, through the safety pressure rupture disks, and into the air from the web of lines around them.

Dey, knowing that a low temperature was important in

slowing this type of chemical reaction, shouted to Qureshi that they should get the refrigeration unit turned on. This system, situated adjacent to the three tanks, was normally left running in order to keep the MIC at a constant, cool temperature. If they could only reduce the temperature of the MIC, he reasoned, the pressure might subside and they would buy some time for figuring out what to do next. Dey ran over to the control panel and turned on the refrigeration system. It had been turned off many weeks before when they began their temporary shutdown of the plant. It never should have been turned off in the first place.

He flipped the starting switch, and the pumps came to life, slowly at first and then faster. The noise from the pumps became louder and louder with each passing second, but something was wrong. These were not normal sounds. Dey stared at the small control panel on the refrigeration unit and suddenly recalled that a maintenance crew had drained the coolant from this refrigeration unit some weeks before and used it in another part of the plant! There wasn't any coolant in the system because management didn't want to spend the money to purchase it. The pumps inside the refrigeration unit were pushing air - - not liquid coolant - - and were now on the verge of tearing themselves apart. He slammed his open hand against the metal control panel in anger, flipped the power switch off to shut down the refrigeration unit, ran back to the MIC tanks, and told Qureshi and the others that the unit was inoperable.

Like the other operators and his boss, Dey had never taken any emergency procedures training. He relied on his knowledge of the plant and his gut analysis of what they could do to solve the problem. He suggested to Qureshi that they pump the MIC in tank 610 into tank 619 which was kept relatively empty. Dey and Qureshi ran to the tank level meter only to find that it showed the tank to be nearly full.[1] That wouldn't work either.

The situation had reached a critical stage, and Dey knew it was time to start up the major safety systems of the plant.

He shouted frantically to Qureshi, "I'm going to turn on the vent gas scrubber."

Qureshi directed the field operators to reroute the gas to the scrubber by repositioning valves just as Dey turned around and ran back towards the control room to activate the system. The safety device, located outside beyond the MIC tanks, looked much like a tall white rocket standing in the middle of the complex. It would neutralize the gas by showering it with a caustic soda as it passed through the tall vertical tank.

Dey burst through the doorway of the building and down the hall into the control room. The air inside was appreciably better than it was outside, and he inhaled deeply as he ran to the panel controlling the vent gas scrubber. He switched the system out of standby mode and turned on the recirculation pump. This would circulate the soda down through the tank and over the MIC gas as it was passed through.

Incredibly, something was wrong again! He reached up and tapped on the face of the flow meter display. This was the instrument that told him how much caustic soda was circulating down through the tank and over the MIC gas. There was no movement on the display. Nothing was flowing![2] Dey

[1]It was determined after the event that the tank *was* nearly empty and the gauge provided a highly inaccurate reading.

[2]A subsequent investigation showed that the soda was likely flowing through the scrubber and that the flow meter had failed to work due to lack of maintenance. Regardless, the scrubber was designed to neutralize a maximum of 190 pounds of gas per hour at 15 psig and a temperature of 95° F. The MIC entered the scrubber at a temperature in excess of 400° F and at a rate of over 40,000 pounds per hour, more than 200 times the design limit.

rushed out the control room once again knowing full well that their options were running out.

Bolting out of the building once again, Dey sprinted past the MIC storage tanks and out toward the tall vent gas scrubber, his lungs burning worse than ever. All of the field operators were standing by waiting for some sign that the system had been activated and the gas was being controlled or neutralized. Dey had not yet reached his co-workers near the scrubber when there was the sound of a loud crack against the backdrop of the screeching safety valves - - it was another rupture disk giving way. He stopped, looked up into the sky toward the source of the noise, and saw the enormous plume of MIC gas spewing out the end of a vertical pipe over 100 feet above his head. It was far worse than before. The chemical brew inside tank 610 was now over 400° F and pushing the gas through the pipes and out the tall vertical pipe at the phenomenal rate of more than 40,000 pounds per hour! There was far more energy than could be contained by the plant's plumbing, and the MIC facility was now nothing less than a runaway machine - - a machine that was spewing tens of thousand of pounds of poison vapor out into the night air.

Day ran to Qureshi, yelling above the noise of the gushing MIC vent 100 feet overhead.

"Let's divert it to the flare tower." He knew they had to act fast now; the plume was forming a massive cloud of gas which had begun to descend upon the plant and the surrounding area.

The flare tower was a near-final line of defense. Gas could be routed into the tall tower and ignited. It would make a high flame in the sky above the plant with the present flow rate of 40,000 pounds per hour, but the gas would be burned off in the fire.

Qureshi issued orders to the nearby field workers to prepare to turn the valves to re-route the flow to the tower where the gas would be channeled and then ignited. Dey followed him as they ran the dozen or so yards around the corner to the tower. Taking his last running step, Dey stared ahead at a large gaping hole in the side of the flare tower.

He screamed out: "Qureshi! It's missing. The connection is missing!"

There was a large round hole on the side of the flare tower where a pipe should have been connected. Dey realized that there was no way to get the MIC gas channeled over to the flare tower to be ignited. Their most foolproof safety system could not be activated. Dey did not know that maintenance workers had removed the corroded connecting pipe a number of weeks before. It had never been replaced.

The plant superintendent, who had just arrived at the facility and joined the group, ordered the alarm system activated. A field worker ran back into the control room to pass the word. The alarms blared, alerting the 120 employees on duty to evacuate, and Dey stood outside in the dark under the billowing cloud of poison gas watching the doors explode open and the silhouettes of workers spilling out into the yard. Some stopped momentarily and looked up at the huge gray plume in the dark sky. Then they rushed headlong to the gate and then north, into the light breeze and away from the cloud that was forming above the rooftops of Bhopal.

There was just one final thing that they could do: turn on the water curtain. This was a system of high pressure water pipes and nozzles that would spray a curtain of water over the plant and, again, neutralize escaping gas. Dey directed Qureshi and those remaining to open the valves to the water lines. The spray nozzles, each like the end of a fireman's hose, were pre-positioned to blanket the entire production facility in a shower of water.

The first pulses of water rushed through the lines and then spurted through the nozzles a few feet. The pressure increased quickly and a blasting shower of rain enveloped the facility. Suman Dey was too exhausted to notice or care that he was soaking wet within a few seconds. He placed his hand above his eyes to shelter them from the pelting spray of water and looked up into the sky. There stood the tall pipe and the spewing plume of deadly MIC. He lowered his hand and rubbed his wet face, fully aware that the massive spray of water was not powerful enough to reach the gas plume. The hole from which the gas streamed was more than 100 feet above the ground and the water curtain reached only 40 feet up into the air - - 60 feet short. He knew he had to leave now if he wanted to stay alive. Dey ran along across the wet gravel toward the gate and then through the dark alleyways of the extended ghetto and to the north. He did not stop until he had reached the isolated mountains well outside the city.

The escaping cloud of methyl isocyanate descended upon the sleeping residents of the densely inhabited neighborhoods surrounding the plant. Some residents awoke with searing eyes and lungs, only to suffocate and die while running through the dark and narrow alleyways between their makeshift homes.

Others suffocated in their sleep, drowning in the fluids excreted by their own lungs. The final count was never known, but at least 2,500 residents of Bhopal died in the early morning hours of December 3, 1984. Tens of thousands more suffered irreparable respiratory and neurological damage. Suman Dey and all of his co-workers at the plant survived.

After their arrest at the airport, Warren Anderson and his colleagues were taken to Union Carbide's comfortable guest house, protected from the mob outside by 50 armed guards. Released six hours later on $2,500 bond, he made a statement to the press. His speech demonstrated his honest concern for the victims, but also revealed his level of understanding of the causes: "..our safety standards in the U.S. are identical to those in India or Brazil or some place else. Same equipment, same design, same everything."

REFERENCES AND NOTES

A Calamity for Union Carbide. (1984). *Time,* December 17, 38.

All the world gasped. (1984). *Time,* December 17, 20.

Bogard, W. (1989). *The Bhopal tragedy: language, logic, and politics in the production of a hazard.* Boulder, Colorado: Westview Press.

Bowonder, B, Kasperson, J. X., and Kasperson, R. E. (1985). Avoiding future Bhopals. *Environment*, 27 (7), 6-37.

Everest, L. (1985). *Behind the poison cloud: Union Carbide's Bhopal massacre*. Chicago: Banner Press.

Graff, G. (1985). Beyond Bhopal: toward a "fail safe" chemical industry. *High Technology*, April, 55-61.

India's Night of Death. (1984). *Time*, December 17, 22-31.

Lepkowski, W. (1985). People of India struggle toward appropriate response to tragedy. *Chemical and Engineering News*, February 11, 16-26.

MacKenzie, D. (1985). Design failings that caused Bhopal disaster. *New Scientist*, March 28, 3-5.

Meshkati, N. (1990). Preventing accidents at oil and chemical plants. *Professional Safety*, November, 15-18.

Meshkati, N. (1991). Integration of workstation, job, and team structure design in complex human-machine systems: a framework. *International Journal of Industrial Ergonomics, 7*, 111-122.

Meshkati, N. (1991). Human factors in large-scale technological systems' accidents: Three Mile Island, Bhopal, Chernobyl. *Industrial Crisis Quarterly*, 5, 133-154.

Morehouse, W. and Arun M. (1986). *The Bhopal tragedy.* New York: Council on International and Public Affairs.

Operating problems cited at Bhopal MIC plant. (1985). *Chemical and Engineering News,* February 11, 31.

Shrivasiava, P. (1987). *Bhopal: Anatomy of a Crisis.* Cambridge, Massachusetts: Ballinger Publishing Company.

Stix, G. (1989). Bhopal: a tragedy in waiting. *IEEE*, June, 47-50.

The major elements of this event are believed to be depicted accurately. The dialogue is fictional but based on published accounts and the unfolding events within the story.

NEVER CRY WOLF

It was a pleasant mid-July morning outside the medium-security womens' prison in Oregon, and Diane Downs stood in the recreation yard, eyeing the two fences that kept her inside. It's funny how you sometimes get the feeling that everything is going to work out exactly like you planned. This was such a time, and Diane decided then and there to make her move. There were no guards watching the yard at the moment, and she knew she could make it over both of the tall fences without too much trouble. It was definitely worth a try. After all, there was only one thing they could do if she got caught: throw her back into jail.

The big breakthrough came the day before when she recognized the guards' reliance on the motion detection system located at the second fence. It wasn't that the security system did not work. No, just the opposite was true; it probably was working *too* well. So well, in fact, that the alarm system, together with the guards' conditioned response to it, would enable her to escape.

The alarm had sounded frequently the past few weeks. All it took to trigger the sensor was a bird or two flying low near the base of the fence. Why, even a good wind rustling the tall weeds had set it off. Hardly a day had gone by in the last week that the alarm had not sounded. And that, of course, was her key.

She took one last look around, walked to the chain link fence, grabbed it with her hands, and put the toe of her right shoe in one of the diamond-shaped openings. Up she went, one hand above the other, until she was at the top, 18 feet up. Getting over the razor wire was a little tricky, but not at all impossible. It might have slowed her down, though, if they had been chasing her. With her blue polyester blouse and faded 501 jeans now clear of the wire, she scaled back down the other side. So far, so good.

She turned and walked calmly across the narrow median toward the second fence. Then, just as she expected, the optical motion detectors discerned her presence. A sensor's electronic circuit closed, signaling the prison alarm. It came on immediately. The deafening blare from the speakers washed over the prison yard.

Unfazed, Diane hopped up onto the second fence and climbed its face. She negotiated the wire at the top, perhaps a little more quickly this time, and scaled down the other side, dropping the final few feet to the ground. There were a few faces looking her way, but they were all stares of awe from fellow inmates on the other side of the fences.

Yes, everything had come together just as she planned. The guards, desensitized by the multiple false alerts of the past weeks, were in no hurry to investigate the alarm. In due time, she knew, they would get around to shutting it off. And, when they did, they would find one less prisoner. With that, Diane Downs, incarcerated for shooting her three children, walked calmly across the grass to freedom.

REFERENCES AND NOTES

Another computer-related prison escape (1987). *ACM SIGSOFT Software Engineering Notes*, vol. 12, no. 4, 7.

Rule, A. (1988). *Small sacrifices: a true story of passion and murder.* New York: Penguin.

LEAP OF FAITH

Captain Michel Asseline was, undeniably, excited about piloting the brand new Airbus A320 for the air show at Mulhouse Habsheim in the Alsace region of France, but he would not in any way let that diminish his professionalism. He could not have looked more competent or composed, standing tall and debonair in his perfectly pressed dark blue Air France uniform near the doorway between the cabin and cockpit. As the first passengers approached the aircraft through the boarding tunnel he watched their faces - - they were all thrilled to be participating and clearly impressed with the beautiful new jet. This was definitely more fun than conducting a daily commercial flight in an old 727.

Two passengers, a boy and his young sister, were among the first to enter the cabin. The children turned to look down the aisle; their excitement over the spotless new transport and the demonstration flight they were about to take was contagious. The boy had his eye on two of the seats in row 8. He and his sister would have a clear view during the low-level flyover at the air show.

Michel Asseline needed to do one more thing prior to getting back to his primary duties as Captain: greet the two off-duty air hostesses who were going to sit in the cockpit with him and First Officer Pierre Mazieres during the flight. They were walking into the aircraft now, and one could not help but note how attractive they both were, especially the brunette. First Officer Mazieres, sitting in the right forward copilot seat, turned around to his left as best he could and exchanged buoyant greetings with the two young women as they entered the cockpit. Captain Asseline stood behind them in the small hallway and watched as First Officer Mazieres told them a little about their revolutionary cockpit instruments and the unmatched capabilities of the A320. The women took their places in the two small jump seats, and Captain Asseline stepped over the center console to maneuver into his position in the forward left-hand seat. He and First Officer Mazieres proceeded with their preflight checks as the remaining passengers boarded the plane under the direction of the purser and three flight attendants. Everyone on board sensed that it was going to be an electrifying ride.

The day's flight plan called for a low-level pass over the grass runway at the Mulhouse Habsheim aerodrome only ten nautical miles away. Earlier that morning they flew the A320 from Charles de Gaulle just east of Paris to the Basle-Mulhouse airport here in eastern France. Now, they would take off, fly at low altitude to the Mulhouse Habsheim aerodrome, drop down to 100 feet over the airstrip, and then pull up, come around, and make the pass again from the opposite direction. Captain Asseline knew that the passengers on board, many of whom were taking their first flight on an aircraft, as well as the air show spectators on the ground at Mulhouse Habsheim, would be most impressed with the capabilities of the aircraft and the low-speed, low-altitude flyover, especially as they ascended out of the maneuver under high thrust.

Neither he nor First Officer Mazieres had participated in an air show previously; but no one doubted Captain Asseline's qualifications to conduct the demonstration. His total flying experience as of that Sunday, June 26, 1988 was 10,463 hours and included type qualifications for the Boeing 707, 727, 737, Airbus A300, A310, and, most recently, the new A320. And he was, after all, head of the A320 training subdivision for Air France and had logged 138 flight hours in A320s. He had also spent more than 150 hours in the A320 simulator at Aeroformation and Thomson and had been consulted by Airbus Industrie during the development of the flight control systems.

At 2:29 in the afternoon, with the passengers aboard and the plane ready to go, Captain Asseline eased away from the gate at Basel-Mulhouse airport and proceeded to the taxiway. At this point First Officer Mazieres asked him to verbally review the flight as they made their way slowly toward the runway. Captain Asseline, obliging the request, stated that he would make the first demonstration flight over the runway at Mulhouse Habsheim aerodrome with the flaps extended to Configuration 3 to provide the extra lift necessary for the low-speed run. They would descend to 100 feet above the grass runway with the landing gear down. The plane would be in a maximum nose-high angle of attack as they flew past the crowd of spectators at the aerodrome. First Officer Mazieres was to maintain level flight while they were over the runway by adjusting thrust. He was also, at the request of Captain Asseline, to select maximum engine power while Captain Asseline executed a climbing turn. They would then ascend to altitude, make a wide loop, and return to the aerodrome for the second pass from the opposite direction.

This maneuver, Captain Asseline knew, could be performed with such confidence and ease in only one airliner in the world: the A320. The plane was a *tour de force* of technological achievement and the clear pacesetter in commercial aviation, years ahead of anything produced by Boeing or McDonnell-Douglas. Although the airframe and overall design of the A320 were exceptional, the most revolutionary aspect of the craft was the entire flight control system, and Michel Asseline was one of its most impassioned advocates. The Airbus was a *fly by wire* aircraft. Shielded electrical lines, not metal cables or vulnerable hydraulic tubes, ran from the cockpit controls to the control mechanisms and wing surfaces.

The use of electricity to transmit control inputs was only half the story. Instead of simply routing the pilot's control inputs directly to the wings and engines, the electrical cables from the instruments fed into a bank of five computers. The computers, in turn, analyzed the inputs from the pilot, compared them to various flight parameters such as speed and altitude, and only then sent the signals to the wings, engines, or other major components. This system provided what had come to be known as computerized "flight protection," a revolutionary safety feature heretofore unavailable in commercial aircraft. If the pilot inadvertently approached a stall by going too slowly or pitching the aircraft up too high - - thus leaving the "flight envelope" - - then the computers, programmed with the laws of flight and the countless characteristics of the aircraft, were capable of increasing engine speed or limiting the pitch of the plane in order to avoid the stall. Also, if the pilot maneuvered the plane violently and exceeded the forces the airframe was designed to take, the computer modified the pilot's input and executed the maneuver but kept the plane within the limits of structural safety. Some within Airbus likened the flight protection system to a guardrail on a mountain road. If the driver lost control of the vehicle on a turn, the guardrail was there to save him from going over the edge of the cliff. Similarly, the computers and the flight protection system kept the plane within the flight envelope and within the structural limits of the aircraft. The design was a textbook case in the use of a computer to watch over the operator and keep him out of trouble.

Nearing the runway, Captain Asseline continued his review for the copilot. For the purpose of the demonstration they would disengage the *autothrottle* and the *alpha floor* function. The autothrottle, conceptually similar to a cruise control on an automobile, was one of the systems that automatically maintained a specified speed regardless of the plane's position. The alpha floor function, one of the flight protection capabilities for which the A320 was so famous, kept the plane from stalling during approaches and takeoffs due to any combination of low speed, angle of attack, or loss of lift.

Captain Asseline knew that the flight protection system activated automatically when the plane pitched upward more than 15 degrees and its altitude was 100 feet or more above the ground. The flight plan called for a steep climb up and out of 100 feet, a climb which would likely exceed 15 degrees. It was therefore necessary to deactivate the automatic protection system so that it did not prohibit him from making the steep climb upward. He also knew that the *alpha floor* stall protection system did not come on automatically at an altitude below 100 feet. Pilots landed planes by reducing speed on the final approach and then flaring the wings just as they were over the runway, inducing a controlled stall. With the loss of lift, the plane settled gently onto the runway. The alpha floor protection system safeguarded the plane from stalling, so unless it was automatically shut off at very low altitude the plane would never be "allowed" to land. To the neophyte it all seemed rather complicated, but the flight protection system made a great deal of sense to anyone who flew the A320.

The plane neared the end of the runway and First Officer Mazieres requested and received approval from air traffic control for an altitude of 1,000 feet above the ground for their brief flight to Mulhouse Habsheim aerodrome. They sat and waited for six minutes as conflicting aircraft traffic passed. The weather was clear and a light breeze blew from the north. Then, at 2:41 p.m., they received clearance for takeoff, pushed the throttles forward, and started the roll down the runway. The big A320 reached a speed of 153 knots and Captain Asseline pulled back on the side controller to head the plane up into the air on a heading of 155 degrees. They gained a bit of altitude, raised the landing gear, and set the flaps to position 1. Captain Asseline initiated a smooth right turn and held the plane banked at 34 degrees for one minute until he attained a heading of 010 degrees, slightly off of north. They completed the turn and leveled off 1,000 feet above the ground. Their speed was now 200 knots. He disconnected the autothrottle so he, not the computer, had full control over the engines. They would reach the aerodrome in only two or three minutes and make the low flight in front of the air show crowd.

Captain Asseline turned the plane slightly to the left to a heading of 340 degrees, the direction of Mulhouse Habsheim, and started looking for the small airport. Trees covered most of the terrain, and from the low altitude it was difficult to see any distinguishing features in the sea of green below. His concern about finding the aerodrome was intensifying when, finally, with only 5.5 nautical miles to go, he saw what must be the aerodrome and initiated a slow descent at a rate of 300 feet per minute by moving his side controller forward and pitching the nose down slightly. The delay in locating the little airport left them closer than he had planned. Their final descent would have to be a little steep, but this was not yet fully apparent.

First Officer Mazieres called the tower at Mulhouse Habsheim aerodrome to tell them that they were approaching. "We are coming into view." A few moments later he radioed again: "We're going in for the low-altitude flyover."

The controller at the tower responded by providing altimeter information, which both pilots repeated in order to verify that they were operating with the correct altitude settings, something that would be most important due to their planned 100 foot altitude during the flyover. They checked their altitude instruments one last time and everything matched up.

During all of this, Captain Asseline continued to scan the terrain ahead and study the airport. At last he could make out the grass runway and the crowd of people just to the west, in front of the hangers. He set the engines throttles in the open descent mode, thereby placing the engines in idle so that the big jet would reduce speed and descend. The A320 began to slow at a rate of 1 knot per second. He extended the landing gear and, five seconds later, extended the flaps from position 1 to position 2. Eighteen seconds later he extended the flaps to position 3, and their speed was now down to 177 knots. They were 500 feet above the ground and on a heading of 344 degrees.

About 15 seconds later their speed had fallen to 155 knots and their altitude was 200 feet. Captain Asseline knew that they were coming down a little steeply, but he also knew that they should reach the clearing of the airport at the correct location because of their low air speed. At 2:45 p.m. and 14 seconds they reached an altitude of 100 feet above the ground.

First Officer Mazieres stated, "OK, you're at 100 feet there, watch, watch...."

He stopped talking mid sentence as the computer-controlled voice altimeter reported their altitude:

"100 feet."

A pulse of adrenaline rushed through his chest and arms. Their air speed was now down to 132 knots. The trees and the grass below were coming up fast. He pulled back gently on the side controller and flared the plane, putting her in a 4 degree nose-up position.

The voice altimeter blurted the altitude once again, only this time it was lower, much lower:

"30 feet."

Finally, Asseline knew they had "bottomed out" and were now in level flight above the ground. Their altitude was low, though, and he pulled back more on the controller, bringing the nose up to an angle of 7 degrees. They flew slowly above the ground, much like a large lumbering bird as it lifts its head and spreads its wings wide in anticipation of its feet touching the earth when landing. The plane continued flying 30 feet above the long grass runway in front of the spectators for another six seconds, its nose pitched into the air.

Suddenly, Copilot Mazieres shouted, "Watch out for the pylons ahead..."

Captain Asseline, looking straight ahead over the raised nose of the plane, responded quickly, "Yeah, yeah, don't worry." He recognized the extent of the situation, but he still felt in control.

Now their air speed had fallen to 122 knots and the voice altimeter called out in rapid succession.

"30 feet."

"30 feet."

"30 feet."

The control tower at the airport slid past the side window and registered in his peripheral vision. It was time to make the "go-around." Without saying anything to Mazieres, he pushed the hand throttles quickly forward through all of the detents, each one clicking loudly as the control passed by, to full thrust. His left hand pulled back on the side controller as far as it would go. The nose of the plane lifted up to 15 degrees.

He waited a moment, then another. A second larger surge of adrenaline shot through his chest. He could not feel any throttle response. Christ! They weren't accelerating!

There was insufficient time in those few brief seconds to realize what was actually going on. All Captain Asseline knew was that the engines were not responding. But unlike his 20 practice runs performed at altitude, this time the entire envelope protection system was shut down. He had disengaged the autothrottle system earlier so that he could "power out" in the steep climb. Furthermore, the throttles had been placed in idle during their descent, and the engine turbines were now spinning slowly and providing little thrust. Their forward motion was slowing and the plane had lost its reserves of kinetic and potential energy. It could not gain altitude without thrust, and that would not come until the turbines were up to speed. All of his other demonstrations of the powered maneuver had been conducted as climbs out of level flight at a consistent speed - - a condition in which the turbines were already spinning. Unlike the other times, the turbines now had no inertia and it would take seconds to get them turning.

The A320, meanwhile, continued to lumber along just above the runway at a height of about 30 feet and a nose-up angle of attack of about 15 degrees. Asseline's grip on the side controller in his left hand tightened. He had it pulled back as far as it would go. His right hand clinched the throttle levers, which he had pushed all the way forward. It was difficult for him to see out the windscreen because the nose was up so high, and he knew they were approaching the edge of the forest at the end of the runway which must be only a few hundred feet away by now. There still was not any thrust.

Four seconds after he had pushed the throttles forward, just as the turbines were getting up to speed and starting to provide thrust, he heard Mazieres shout something, and a rapid succession of loud cracks reverberated through the plane as the underside of the tail hit the tops of the trees. The forward motion of the A320 slowed immediately, and the plane settled through the top of the dense forest. Asseline felt the straps of his harness pressing tight against his chest as they crashed downward through the trees. In quick succession the two big engines under the wings sucked in tons of debris and stopped spinning, the right wing broke off as the plane hit the trunks of the larger trees, and thousands of gallons of jet fuel sprayed forward through the air. They slammed into the ground and his head and arms snapped forward uncontrollably; he felt his head hit something hard and then bounce back. The motion of the plane finally stopped and he sat stunned for two or three seconds, yet fully aware of the magnitude of what had just happened. The impact had been horrific, but they were still alive and the plane was intact. Then the spilled jet fuel surrounding the plane exploded into flames.

He reached over and activated the evacuation alarm, which did not come on, and then unbuckled himself from his seat. The two off-duty air hostesses sitting behind jumped from their seats and headed for the door at the back of the cockpit. One of them slammed open the door to the cabin and Asseline saw that the fuselage was still largely intact, although flames were entering the cabin through broken windows near rows 8 and 9. The floor was cracked wide open a few rows behind that and the flames were beginning to come up through the aisle. One of the off-duty air hostesses and the purser pushed on the forward left door where the fire was not yet burning. It was blocked by the tree branches outside. Suddenly, the door opened unexpectedly and the purser and air hostesses fell out of the plane onto the broken trees and bushes. The evacuation slide then unfolded below the doorway and inflated on top of them. Panic-stricken passengers rushed forward toward the doorway and jumped in the escape slide, only to pile up half-way down because of overlying tree branches. Others jumped from the doorway to the ground, but found themselves unable to move aside due to the broken, dense vegetation. Realizing what was happening, the flight attendant and a passenger at the doorway stopped the others from exiting while the purser and hostess outside the plane cleared the branches from the escape slide. The flight attendant helped the passengers by the door escape into the slide and then, overcome by the smoke, she too jumped out the door.

A second group of passengers, some screaming and on fire, rushed forward from the midsection through the smoke and flames, pushing another flight attendant along in front of them. She assumed a position next to the door and directed their evacuation. By this time Asseline had removed the injured Mazieres and taken him to the slide. He then turned to the remaining flight attendant and ordered her to leave. Knowing there were still passengers in the burning plane, he raced back into the cockpit, pulled out his antismoke mask, and ran back to the forward section of the cabin. The heat was now intense and the cabin was completely full of smoke. Overcome by the fumes and heat, he backed up and jumped out the door. Asseline, all of the other members of the crew, and the passengers who had left out the front door ran from the plane as it was enveloped in flames. They met up with other passengers who had managed to evacuate out the left rear door in much the same manner as they had exited the forward door.

As they stood in the woods and watched the firefighters undertake the futile battle to extinguish the blaze, a young handicapped boy in the fourth row remained in his seat, unable to move. Another child, a little girl located in seat 8C, was trapped, unable to open her seat belt. Her brother, seated next to her, attempted to help, but had been carried away by the second rush of people headed for the forward exit. A woman traveling in seat 10B was one of the last ready to leave the burning plane at the forward door. Hearing the cries of the little girl whom she did not know, she ran back into the burning cabin. Her remains were found next to those of the little girl.

The tail section containing the black-box flight and voice recorders was the only substantial part of the plane remaining after the intense fire. On June 27th, just one day after the crash, the *Generale de l'Aviation Civile,* a branch of the French Government responsible for civil aviation, issued an initial report on the accident, an action that many in the aviation community looked upon with great skepticism due to its quickness and tone. The A320, the report said, performed flawlessly and precisely according to its design. "We have no evidence at the moment that puts the proper functioning of the aircraft or its systems in doubt, or raises questions about the A320's safety when the aircraft is flown normally." The pilots were completely responsible for the crash, according to the government report.

On November 30, 1989, more than one year after the accident at Mulhouse Habsheim aerodrome, the final results of the *French Ministry of Planning, Housing, Transportation and Maritime Affairs Investigation Commission* were made public. In addition to faulting the organizers for the whole stunt, the commission concluded that the crew had not properly prepared for the maneuver. Their planning for the task was incomplete, and the level of coordination between the two pilots was not clearly understood by both. The aerodrome was located late, and the overflight was not conducted according to the original flight plan due to the rush of events. This resulted in the flight parameters not being stabilized during the flyover and the engines remaining in idle and, subsequently, slow to respond. Captain Asseline, the commission concluded, was influenced by the holiday atmosphere associated with the air show and his ardent support of the A320 technology and his desire to defend the aircraft. He was also likely influenced by the presence of the female hostesses in the cockpit.

Lastly, it was determined that the very system designed to increase the safety of the craft, the flight protection system, inspired overconfidence in the captain and was a probable cause of the accident. His training had emphasized all of the protections provided by the A320 and contributed to his lack of caution in performing a maneuver well beyond the limits of safety. He was lulled into a false sense of security without fully realizing that his performance with the A320 was in large part due to the presence of the flight protection system. Instead of treating the flight protection system like the safety guardrail on the winding mountain road, he employed it to define the limits of aircraft flight. It was as if he used the guardrail to negotiate the curve, not treating it as a protective barrier placed there in the event that he lost control. He had turned the flight protection system off for the maneuver and thereby had to fly by the standard rules of flight - - something for which he may not have been entirely prepared.

On February 14, 1990, 19 months after the incident at Mulhouse Habsheim aerodrome and ten weeks after the release of the report by the Commission, an Indian Airlines Airbus A320 crashed on approach at Bangalor Airport, killing 94 of the 146 persons on board, including the two pilots. The investigation of the crash determined that one of the crew members had disconnected the autothrottle and selected the *idle open descent mode* during the final phase of descent to the airport. The system, designed to be used only at high altitude during descent with reduction in speed, was activated during approach. Unknown by the crew, the engines slowed to idle speed and the plane, flying at a minimum speed and maximum angle of attack, began to stall at an altitude of 140 feet. The pilot applied full throttle and the engines began to respond some seconds later as they overcame the inertia of the idle turbines, but not before the plane slammed into the ground short of the runway. The investigators concluded that crew overconfidence played a significant role in the accident.

The following month, Airbus Industrie conducted a special A320 operators meeting in Toulouse to discuss the development of measures to counteract an "overconfidence syndrome" believed to exist among A320 crews. In addition to changing the training strategy, plans to upgrade the "alpha floor protection mode into an approach protection mode to prevent descents without sufficient thrust" were discussed.

REFERENCES AND NOTES

Airbus may add to A320 safeguards, act to counter crew "over-confidence." (1990). *Aviation Week & Space Technology*, April 30, 59.

Alizart, R. (1989). Electric airliners: letters to the editor. *Science*, 244, August 11.

Cabin crew evacuated A320 passengers until smoke, flames became too intense. (1990). *Aviation Week & Space Technology*, June 25, 98-99.

Chisholm, P. (1988). Troubled skies for the Airbus. *MacLean's*, July 18, 36.

Commission finds flight preparations not in accordance with Air France rules. (1990). *Aviation Week & Space Technology*, July 23, 90-93.

Commission proposes more preparation, special training for airshow flyovers. (1990). *Aviation Week & Space Technology*, July 30, 90-93.

Flight recorders detail A320's engine, control settings for air show flyby. (1990). *Aviation Week & Space Technology*, June 11, 78-79.

French Report details 1988 crash of A320 following air show flyby. (1990). *Aviation Week & Space Technology,* June 4, 107-109.

Gavaghan, H. and Watts, S. (1988). Crash fuels worries about computers aboard Airbus. *New Scientist*, June 30, 36.

Gunston, B. (1988). *Airbus*. London: Osprey Publishing Limited.

Lenorovitz, J. M. (1990). Indian A320 crash probe data show crew improperly configured aircraft. *Aviation Week & Space Technology,* June 25, 84-85.

Lenorovitz, J. M. (1988). A320 crash inquiry finds no aircraft technical faults. *Aviation Week & Space Technology,* August 8, 28-31.

Lenorovitz, J. M. (1988). A320 crash investigation centers on crew's judgment during flyby. *Aviation Week & Space Technology,* July 4, 28-29.

O'Hare, D. and Roscoe, S. (1990). *Flightdeck performance: the human factor*. Iowa State University Press.

Post-crash fire destroyed A320 following plunge through trees. (1990). *Aviation Week & Space Technology,* June 18, 99-93.

Simulator, actual flights reproduced A320's maneuvers, engine parameters. (1990). *Aviation Week & Space Technology,* July 9, 60-62.

Smyth, R. (1988). The ugly search for error. *Alexandria Journal,* July 1.

Waldrop, M. M. (1989). Flying the electric skies: Who's minding the cockpit? *Science*, 244, 1532-1534.

Wiener, E. L. (1989). Reflections on human error: matters of life and death. *Proceedings of the Human Factors Society 33rd Annual Meeting*, 1-7.

All of the dialogue in this story is based on the record from the aircraft's flight recorders.

THE WIZARDS OF WALL STREET

"Stocks ended mixed after a late burst of computer-driven selling erased gains made earlier in the day...

The Dow vacillated much of the day, only to drop by a brisk 15 points in the final 10 minutes. Traders attributed the sudden fall to heavy computerized program selling..."

The Los Angeles Times,
Thursday, March 26, 1992
Business Section - -
Financial Markets

It was Wednesday, March 25, 1992, when Michael Baltieri was handed the transaction order by one of the senior traders at Salomon Brothers in Manhattan. Michael laid the slip of paper down on his desk and looked up and surveyed the controlled state of chaos in the expansive room. The trading floor at

Salomon was abuzz with the usual contingent of traders, brokers, executives, managers, and clerks. These people, like their counterparts at Merrill Lynch, Shearson, and the other big brokerage houses, would have bought and sold billions of dollars worth of securities for their clients around the world by day's end. It would have been nice to sit back and watch the ceaseless activity on the trading floor for a few minutes; but there was no time for that. Michael was a clerk at Salomon, and his job was to enter "buy and sell" orders into the computer at his cramped workstation. The "sell" order that had just been handed to him had to be processed before the day was out. He had to hurry to get under the deadline.[1]

Michael took a quick look at his wrist watch: it was 3:50 in the afternoon, just 10 minutes away from the closing bell at the New York Stock Exchange down the street. The noise level on the trading floor at Salomon had increased appreciably during the last few minutes. It was always like this near the end of the day as buyers tried to complete their transactions before trading stopped at the Exchange. It seemed that everyone wanted to get something in just under the wire. The activity on the trading floor would grow into a deafening crescendo as brokers issued their final transactions to take advantage of the current trends and projections for the next day's stock values.

Michael Baltieri was well aware of the huge increase in last minute trading in the past few years. After all, he was one of those guys down in the trenches who had to deal with it every

[1]The actual manner in which the "sell" order was given to the clerk could not be verified. However, a "buy or sell order" is typically transferred to a clerk with a hand-written ticket.

day. The new computer technology had been brought in to take over many of the most demanding trading tasks. But when all was said and done, the powerful computers had increased, rather than decreased, the workload for countless people like Michael. Brokers had also created new and inventive things to do with the computer networks, and computerized program trading was one of the most far reaching.

The computer terminal in front of Michael was the most visible and symbolic part of the whole system. It was connected to Salomon's Sun Microsystems computer system which, in turn, was linked to the Big Board at the NYSE via the SuperDot communication and processing system. Traders and clerks like Michael could initiate the purchase or sale of millions of dollars worth of stock - - or even millions of shares of stock - - in an instant. This program trading system could constantly track the value of shares and buy or sell huge amounts of stock when predefined "lows" or "highs" were reached. The system could execute a transaction in the last minutes of the day in anticipation of rising or falling values at the opening bell the next morning. It could even execute a "buy" or "sell" order expressed only in terms of dollars by distributing the funds across a broad range of predefined stocks, such as the S&P 500. Big "market on close" orders, made possible by the growing power and speed of the computer systems on Wall Street, had made these last few minutes of the day unbelievably frenzied.

Michael knew that people, if given the opportunity, would conduct these massive program trades until the last possible second. He was also aware of the tremendous amount of responsibility it placed on a single person. In this case, *he* would shoulder the burden.

Michael looked down at the order. It was from one of the big clients and had been sent electronically to Salomon. He was to "sell 11 million" of stock, according to the order. After entering assorted information on the client and the transaction on the keyboard, he toggled the cursor over to the "share amounts" field to make the primary entry for the transaction. He read the paper again and understood it to mean that he was to sell 11 million shares of stock. In actuality, however, the order was telling him to sell 11 million *dollars* of stock - - not 11 million *shares*. It was an important distinction because the powerful computer at his fingertips was capable of handling both types of transactions.

There were six minutes left until the closing bell. The noise level on the trading floor was slowly building to a thunderous roar. In a few minutes it would peak and then vanish. Everything on the screen looked fine, so Michael entered 11,000,000 into the field.

Michael was now on the brink of instructing the Big Board SuperDot computer system at the Exchange to sell 11 million *shares of stock* for Salomon - - not 11 million *dollars* worth of stock as the client and the trader who had given him the order intended. The value of the 11 million shares was vastly greater than 11 million dollars. Now it was 3:55 p.m., and in less than a minute the New York Stock Exchange and trading floors throughout Wall Street would be thrown into a near state of panic.

But there was still hope of catching the simple mistake and stopping the sale. Salomon policy required a second clerk to double-check the order before it was finally sent through the company's computers and over to the Exchange. At that particular busy moment, however, no one was available to make the back-up check. The closing bell was approaching fast and Michael knew that the computers might take a few minutes to register the trade during this hectic time of the day.[2] This order had to be placed now!

He set the process in motion by hitting the entry key with that little bit of added emphasis one has when completing a simple yet very important task. Salomon's computers began to churn away and identify the 11 million shares of stock to be traded. The operation was complete within the blink of an eye. Eleven million shares of common stock were to be sold - - 11 million shares of stock worth nearly *500 million dollars!* And it was all because of a misunderstanding about a half-dozen written words.[3]

The computer lined-up the electronic selling instructions and began to route them down the dedicated transmission lines to the New York Stock Exchange. From there the "sell" orders were flashed across the Exchange's worldwide network and back to the Wall Street brokerages. Salomon's computers tagged the transaction as it was underway, but it was too late.

Pandemonium broke out on the trading floor of the NYSE almost immediately at 3:56 p.m. Within less than a minute, the

[2]Published accounts are unclear as to why a second clerk did not double-check the transaction.

[3]Published accounts of this event suggest that the clerk misinterpreted the trade order, intentionally selling 11 million shares instead of 11 million dollars worth of shares. However, it is possible that the clerk simply (and mistakenly) entered 11,000,000 into the wrong field on the display. Descriptions of this aspect of the event are not entirely clear, and an agent of the Securities and Exchanged Commission declined to provide any information on this or related findings pertaining to their investigation of the errant trade.

Dow Jones Industrial Average began to plummet as traders all over the City issued orders to sell. Something very, very big had obviously happened. No one wanted to be caught short during a big drop in the market after a huge "market on close" sale such as this.

Back on the trading floor at Salomon, Michael Baltieri sat stunned as the commotion around him grew. An errant transaction had been made, and it became apparent to everyone on the floor within a minute. It was not immediately apparent to the traders that Michael was responsible, and it did not matter. Rapid damage control was infinitely more important. This was a mistake of alarming magnitude, and the firm had just minutes to somehow tackle the problem.

Senior traders burst from their offices. Shouting matches erupted as managers tried to find out what was going on and, more importantly, what to do. Michael sat petrified, watching as one senior manager ran off to make an emergency call to the Securities and Exchange Commission and as others tried to notify the Big Board of the critical mistake. More Salomon traders hedged their bets and mitigated much of the financial impact of the error by quickly buying stock-index futures on Chicago futures exchanges. At least this way they could cover themselves from some of their losses by betting on stock prices falling or rising. It was akin to knowing the next card to be dealt during a game of blackjack. The Salomon traders knew which way things were headed, and they might cover some of their losses in the process.[4]

---

[4]This account of Salomon's actions immediately following the error is based on information provided in the *Wall Street Journal* on March 27, 1992.

At the closing bell at 4:00 p.m., just five minutes after it all began, the Dow Jones Industrial Average had fallen 15 points, erasing a rally that was underway just as the error was made. Fortunately, the day closed out with a marginal loss of just 1.57 points at 3,259.39. Salomon Brothers Inc. was able to recover all of its losses during the next few trading sessions courtesy of cooperative SEC regulators, brokers, and Exchange managers. They immediately instituted both procedural and design modifications to minimize the chances of a repeat occurrence.

Yet everyone recognized that a major financial disaster could have occurred had the error been made earlier in the day when traders had more time to respond. Accordingly, the Securities and Exchange Commission immediately undertook an investigation into the affair and began to review its policies covering program trading operations and large "sale on close" orders. Thanks to understanding managers and Salomon's recognition of the vulnerability of systems and human interfaces, Michael Baltieri did not lose his job as a clerk on Wall Street.

REFERENCES AND NOTES

Blue chips slip on Salomon goof: broker admits error on programmed trade (1992). *San Francisco Chronicle*, March 26, B3.

Clerk's error stirs worry on street: SEC vows response; big board presses for trading changes (1992). *The Wall Street Journal*, March 27, C1.

Dow dips 1.57 on economic news (1992). *The Los Angeles Times,* March 26, D4.

Dow index picks up 8, rebounding from flub (1992). *The San Francisco Chronicle,* March 27, B3.

Dow rebounds from Salomon goof, rises 8.28 (1992). *The Los Angeles Times,* March 27, D4.

Dreman, D. (1989). A Neanderthal speaking: After the crash of 1929 Congress had the good sense to move to prevent a recurrence. Why hasn't that happened today? *Forbes,* December 25, 164.

Error revives Wall St. debate (1992). *The Los Angeles Times,* March 27, D4.

A pseudonym has been used for the main character in this story. Any similarity to his real name is purely coincidental. Furthermore, the thoughts and observations attributed to this character are fictitious, but are believed to be representative of the setting and circumstances.

GENIE IN THE BOTTLE

The fireman at the Central Facilities A.E.C. Fire Department looked up suddenly from his desk and stared out the window at the moonlit expanse of dirt and sage brush. He would have sworn that something had moved outside, but nothing looked unusual. Deciding he was mistaken, he turned his attention back to his paperwork. Just then the alarm sounded throughout the building. He and five other firefighters, an assistant fire chief, and a security officer were in their emergency vehicles within a minute and speeding across the plains of the National Reactor Testing Station in Idaho toward the SL-1 (Stationary Low Power Reactor) area eight miles down the road. The time was 9:01 in the evening on January 3, 1961, and the worst nuclear reactor accident in U.S. history had just occurred.

Nine minutes later, the trucks pulled off Fillmore Boulevard and stopped at the front gate of the SL-1 area. Like other facilities on the 900-square-mile testing station, SL-1 consisted of a few acres of barren land and half a dozen buildings, all surrounded by a chain link fence and guardhouse. The emergency crew knew they were dealing with a fire or radiation release, or perhaps both. It was also possible that this was a false alarm, but it seemed unlikely given the number of alarms and the late hour. There was every reason to assume the worst had happened, and the alarms suggested that there was a major problem behind the fence guarding the SL-1 nuclear reactor.

Having unlocked the gate, they drove their fire truck a few feet past the unmanned guardhouse and parked it between the administration building on their left and the support facilities building on their right. The tall white circular structure just beyond the support facilities building on the right contained the SL-1 reactor under development for the Army. Everything looked calm and quiet, although it was dark and much too cold out to expect anyone to be walking around outside. Every member of the crew knew that there would be very few, if any, people at the SL-1 facility at this hour, and anyone who was there would most likely be working on the reactor. It had been shut down for two weeks, but was scheduled to be up and running in the next few days.

After discussing the plan of attack, two firemen donned self-contained breathing equipment, picked up radiation survey meters, and ran the short distance in the dark to the south door of the administration building on the left. They unlocked the door and entered, finding the building quiet and unoccupied. The firemen were relieved to find no one inside, yet horrified when they looked down and read the displays on their radiation survey meters. Both showed 25 rems per hour, the highest reading the instruments were capable of and many hundreds of times above the normal background radiation. Knowing that the search had to continue, but knowing also that their own lives were in grave danger, they retreated from the building and ran back to the vehicles and crew near the front gate to discuss the situation and the next step.

Another team of two donned self-contained breathing units within minutes and entered the support facilities building on the right. Finding no one inside in the halls and offices, they raced to the far end and to the foot of the stair leading up to the operating floor of the cylindrical reactor building. Again, their meters registered at the limit of the scales (25 rems per hour) and they retreated from the building, checking rooms and hallways quickly once again for any victims or signs of fire. They hurriedly made their way out the door and to the front gate to discuss the situation with the growing crowd of firemen and radiation experts.

By 9:35 p.m. other facilities on the huge station were contacted by radio. The crew knew now that three men - - two Army and one Navy technician - - were working the 4:00 p.m. to midnight shift at SL-1. They were doing maintenance work on the reactor. These three servicemen had not shown up anywhere else, so it was reasonable to conclude that they were still inside. The emergency crew had a major problem on their hands if the servicemen were inside the reactor building, however, because the steps leading up to the room were within a field of deadly radiation.

Word of the disaster spread fast throughout the complex. Two more health physicists arrived at SL-1 already suited in radiation gear and ready to go inside. It was decided that one of these two technicians with a 500 rem per hour survey meter and two firemen would enter the support facilities building and, if possible, the adjacent reactor building. The three checked their equipment one last time and approached the near end of the structure. Once inside they made their way down the hallway to the base of the stairs leading up to the work level of the reactor. With each step up the stairs, the meter needle crept up the scale, and as they took the last steps up to the dark landing it hit 200 rems per hour, a fatal level. Like the two teams that preceded them, they knew it was far too dangerous to continue, and the team withdrew quickly, running down the hallway and back out to the meeting place near the gate.

Three entries into the buildings had now been made, each one uncovering worse evidence than the previous excursion, and each time failing to find the men who were certainly inside somewhere. There was no choice about what had to be done. Someone must go in and locate the three servicemen.

With explicit approval from the Atomic Energy Commission manager who was now at the facility, one fireman and one radiation technician volunteered to enter the reactor building, go up to the top of the stairs, and quickly survey the round room at the top of the reactor. The two men suited up in front of the headlights of the vehicles congregating at the gate. They grabbed the high-scale meter and raced into and through the building and up the stairs one floor to the working level at the top of the reactor. Peeking around the doorway, it was readily apparent that there had been an explosion of some kind. Debris lay scattered all over the circular floor. But there was no sign of the three men. The technician stared down at his meter, aghast at the measurement. The needle was at the limit of the 500 rems/hour scale. They left quickly and joined the others at the front gate.

With the picture becoming clearer by the minute, it was obvious to everyone that even more drastic action had to be taken. By 10:35 p.m. the SL-1 Plant Health Physicist and Supervisor for Plant Operations from the civilian contracting company in charge of running the reactor arrived and volunteered to suit up and enter the operating floor of the reactor building. Minutes later, they ran up the stairs and to the landing, but this time entered the room, stepping over the piles of dusty debris and gravel-like material covering everything. Before them lay evidence of something that was never supposed to happen, something that *they* said could not possibly occur. Yet it had.

Two minutes later they ran back out for help and returned with two additional contractor personnel, a health physicist from the Atomic Energy Commission, and a stretcher. They were joined by two additional technicians at 10:48 p.m. What they saw during those brief minutes inside would haunt them for the rest of their lives.

Earlier that day at 4:00 in the afternoon, three young servicemen, Navy Construction Electrician first class Richard Legg, Army Specialist fifth class John Byrnes III, and Army Specialist fourth class Richard McKinley, assembled at the SL-1 reactor to relieve the day shift. Their job this evening was to complete the installation of the 44 cobalt-aluminium flux-measuring rods in coolant channels inside the reactor. These flux rods would help the engineers measure the distribution of power within the reactor core when they operated at higher power levels later in the month. This was all a continuation of test and evaluation activities that had been underway since the SL-1 reactor was first fired up over two years before. During the day there were usually about 60 people working at the SL-1 facility, but this evening it would be just Legg, Byrnes, and

McKinley behind the locked front gate.

Throughout the past year the SL-1 had proven itself to be a powerful yet compact source of power. Except for an occasional problem with a "sticky" control rod, the reactor had demonstrated its ability to operate consistently at a thermal power of 3000 kw. It was being developed for the Army, which planned to use it to provide electricity and heat for remote and secretive military installations in the arctic. The reactor core was just over two feet tall and three feet wide. This cylindrical device contained vertical tubes of enriched uranium. When active and in a critical state, neutrons shot throughout the core, struck uranium atoms, and released other neutrons to do the same. It was a rapidly cascading nuclear process that generated a tremendous amount of heat. Water was turned to steam when pumped through the core and around these fissioning columns of enriched uranium. The steam, in turn, would be used to turn a turbine to make electricity or provide heat to counter cold arctic conditions. But that would all be for the production version. For now, they needed to continue with the engineering tests on this prototype.

The highly radioactive core of the SL-1 sat near the bottom of another cylinder called the pressure vessel, which was really nothing more than a very sophisticated, heavy-duty boiler about 15 feet tall and 5 feet wide. The bottom of the pressure vessel rested a few feet above the ground, and the top, including the lid or head, was about 17 feet above that. The "working level" was at the top, which was formed by thick, densely-packed gravel and concrete surrounding the pressure vessel nearly 20 feet down to the ground. All of this sat inside the white steel reactor containment tank, 48 feet tall and 38 feet in diameter. To get to the working level of the reactor (ie., the top of the pressure vessel), one had to simply walk up the stairs at the far end of the facilities building and onto the floor inside the containment tank.

In the center of this circular room, then, was the top or "head" of the reactor, about five feet in diameter and looking much like a huge steel hubcap lying on the floor.

The first shift of the day on January 3, 1961, the one that worked from midnight until 8:00 a.m., had removed the nine drive rod housings atop the reactor head. Each housing, looking roughly like a church bell, had two purposes. First, it protected an elaborate collection of drive mechanisms used to insert and withdraw a rod from the reactor underneath. Second, each housing covered a hole, or nozzle, leading down into the reactor. The second shift had completed the job of installing the 44 flux-measuring rods, and they did this by inserting them down through the nine nozzles and into the slots within the reactor core about a dozen feet below. Legg, Byrnes, and McKinley were instructed to put the housings and assorted components back together so that the reactor could be re-pressurized and fired up the next day. This was everyone's first day of work since the reactor had been shut down two weeks before at the beginning of the holidays.

Without a great deal of discussion about the work that lay before them, the three servicemen went about the business of reassembling the rod drive housings over the nozzles at the top of the SL-1 reactor. Legg, a skilled Navy electrician and mechanic, did much of the detailed assembly work, with Byrnes and McKinley providing the much-needed extra hands to hold components and tools. The tricky part was getting all of the drive mechanisms lined up perfectly so that each control rod traveled upward smoothly through its nozzle and drive housing. The control rods were a critical component and were used to regulate the power output of the core, much like an accelerator pedal is used to modulate the power from a car's engine. When running, the reactor was controlled from a nearby control room from which an operator remotely raised and lowered the control rods via the drive housings and nearby electric motor. All of these components had been disassembled by the earlier shifts, and Legg, Byrnes, and McKinley were well on their way to having it all put back together.

There were nine control rods in the SL-1 reactor, one for each nozzle and drive housing. Each rod looked something like an oar from a row boat, with a long straight handle connected to a long and wide blade. The blade was the critical part and was designed to slip down into slots between the vertical tubes of uranium fuel within the core. Four of the rods had "T"-shaped blades and were used on each of the four sides to form a kind of wall around the stacks of uranium fuel. The five interior rods, one directly in the center and the others spaced out toward the four corners, each had a "+"-shaped blade. When inserted down into the core, the blades absorbed the neutrons darting through the reactor and stopped the nuclear chain reaction. And as the blades (attached to the rods) were raised out of the core, individually or together, the neutrons shot freely toward other uranium atoms and began the process of nuclear fission. A rapid flow of water through the core carried away the tremendous heat generated during this process.

To the three servicemen, the SL-1 reactor was not imposing or threatening, although they each had tremendous respect for its capabilities and compactness, and they were well versed on the precautions of working with radioactive material. What they did not fully realize, however, was that each of the rods on the SL-1 reactor, especially the center rod, was a "high-worth" rod. Unlike most larger reactors containing dozens and even hundreds of independent control rods, each of the nine SL-1 rods regulated a highly significant proportion of the nuclear reaction within the compact core. When work on the SL-1 first began in 1955, it was recognized, theoretically at least, that the core could go "critical" with the removal of a single rod. It was reasoned, however, that the reactor would run nonstop for three years between refueling and that water would always circulate through the core to carry away the heat. No one should ever have the need to remove a rod during operation. But for now, no nuclear reaction was taking place because all of the control rods were inserted into the reactor, their blades well inside the core.

Richard Legg knelt down on top of the circular pressure vessel head on the floor and sorted the parts and components for the center rod drive housing. Byrnes and McKinley stood nearby, Byrnes with his right boot up on the large metal head. The time was half past eight, and they were attempting to assemble another drive housing, this time the one directly in the center. The reactor operating instructions for the procedure were as follows:

FOR DISASSEMBLY
OF THE DRIVE HOUSING...

9. Secure special tool CRT No. 1 on top of rack and raise rod not more than 4 inches. Secure C-clamp to rack at top of spring housing.

10. Remove special tool CRT No. 1 from rack and remove slotted nut and washer.

11. Secure special tool CRT No. 1 to top of rack and remove C-clamp, then lower control rod until the gripper knob located at upper end of fuel element makes contact with the core shroud.

FOR REASSEMBLY
OF THE DRIVE HOUSING...

1. Assembly of the rod drive mechanism, replacement of concrete blocks and installations of motor clutch assembly are *the reverse of disassembly....*

Legg, quite used to dealing with inane instructions like this from his Navy work, proceeded with the reassembly, with Byrnes and McKinley helping out. First, according to their interpretation of the instructions, they had to lift up the rod so that other parts could be slipped into place. Because the electric drive gear was disassembled, raising the rod was accomplished by attaching the "T"-shaped handle and lifting up with about 80 lbs. of force. Once raised four inches, they had to place the C-clamp on the rod to hold it in place and prevent it from slipping back down until they were ready. Legg grabbed the center rod and lifted it up about four inches, but it was obvious that this would not provide enough space to replace the washer and nut on the gear stud.[1] This being the case, they raised the rod just under seven inches above the full-down position. The C-clamp was brought in and tightened into place around the rod to keep it from falling back down. With that accomplished, they removed the special CRT tool still attached to the assembly,

[1]They were, in fact, quite correct. Although only four inches of clearance were required for disassembly, the rod had to be raised more than five and a half inches for reassembly. In either case, there were no scales, markings, or mechanical stops on the components to show the level of the rod. The operator(s) had to estimate and not exceed the distance. There were no other warnings in the procedures regarding the consequences of raising the rod beyond the limit specified.

dropped the scram stop washer on the post, threaded the big nut back down on the gear rack stud, and replaced the special tool on the stud above the nut.

It was now 9:00 p.m., and they were just about finished with the central number 5 drive head. All that remained to be done was remove the C-clamp and reinsert the rod back down fully into the core. The C-clamp was holding the rod up, and it looked like it would be easier to remove if they could lift the control rod up just a bit. Legg could lift up the rod while Byrnes and McKinley bent down and removed the C-clamp. But, when Legg squatted down and grabbed onto the handle and pulled up, it would not budge. He let go, wiped his hands on his pants, grabbed the handle, and took a deep breath, ready to try it again. Legg pulled up as hard as he could for about three seconds. The veins in his neck bulged and his face turned red. But the rod still would not move. As had happened 40 times during the last two months, they had a stuck rod, the blade obviously bound up by something down in the core.

There was only one thing to do. They could certainly dislodge the rod with twice the strength, so Byrnes bent down and grabbed one side of the handle to help out. He and Legg, their knees and backs bent, positioned their feet between the round nozzles on the head and prepared to pull up on the handle. McKinley stood by, ready to remove the C-clamp. Byrnes counted to three out loud and together they pulled up on the handle with all of their strength. The control rod blade down below in the core broke loose suddenly and the rod bolted upward.[2] There were no mechanical stops or structures to

---

[2]The investigation into the event determined that all three men were standing on or near the head at the time of the explosion and that one and probably two of them pulled up on the rod in order to dislodge it. Investigators were unable to determine which two men actually pulled up on the rod.

keep the rod from rising too high.

Dislodged from its shroud inside the core, the rod shot up about a foot from its clamped position before Legg and Byrnes could stop pulling and release their grip on the handle. But it was too late. With the blade extracted, there was now a critical mass of uranium down inside the core. One neutron smashed into a nearby uranium atom and dislodged yet another neutron. The process cascaded a million times over in just thousands of a second, and the temperature of the core shot instantly to 3,740° F. Metal components and the rod blade shroud inside vaporized, and half of the fuel elements in the core melted. The standing water in the "shut down" core expanded with the flash of heat and slammed against the side of the reactor, bulging it outward. The long spear-like control rods ejected upward from the open nozzles at the top of the head with a pressure of 35 atmospheres, and the entire pressure vessel then separated from its foundation and connecting water feed lines and rose up from the floor and bashed into the ceiling of the containment building. Unbelievably, it fell back down into the hole from which it came. With much of the core melted or ejected out the nozzles, the fissioning inside stopped abruptly, and all was quiet a scant two seconds after the operators had reached down and begun to lift up the control rod.

At about 10:46 p.m. the Plant Health Physicist and Supervisor for Plant Operations returned to the working floor of the reactor building with the stretcher. With them were two additional contract personnel and another A.E.C. Health Physicist, all suited-up in exposure gear and self-contained breathing equipment. They worked quickly to place a man, who was alive but not recognizable, onto the stretcher and then examined another man nearby on the floor. He was dead. They left the room just as the two additional technicians rushed in to search quickly for the third man who had not yet been found. The man on the stretcher was taken outside, placed in a van with a nurse, and driven out the gate and down the highway to a waiting physician. He died a short time later at 11:15 p.m.

Back inside the reactor building, the two technicians, instructed to spend no more than two minutes inside the deadly radiation field, searched frantically for the third man. They combed the floor and the piles of equipment and tools thrown about during the explosion. Having looked everywhere, it seemed, and their time nearly up, one happened to aim his flashlight upward. There, directly above the reactor head, was a man, his back pinned to the high ceiling, his face and arms hanging down toward the floor. The bottom end of a dart-like control rod protruded down from the center of his chest, its upper end lodged deep into the ceiling behind his back.

EPILOGUE

Six days later on January 9, the third man's body was finally removed from the ceiling of the reactor building by workers with long poles and grappling hooks. He was suspended within a 1,000 rem/hour radiation field, and the removal operation required the use of numerous teams of technicians, each working for just a few minutes inside. Richard Legg, John Byrnes, and Richard McKinley were highly radioactive, even after repeated attempts at decontamination. They were buried in lead coffins. Twenty three additional persons received in excess of 3 rems of radiation exposure during the rescue, some as much as 25 rems.[3] Decontamination and disassembly of the

---

[3]The following individuals were awarded a certificate for heroism for

SL-1 reactor was completed in November, 1962, nearly two years later.

As a consequence of the accident, many reactor designs, especially those placing exceptional responsibility in the hands of a single operator, were modified. There was a notable move away from the use of few high-worth rods to the use of many low-worth rods. Added emphasis was given to procedural review for reactor operations, and suggestions were put forth that future designs not require manual rod removal for basic maintenance work. Mechanical constraints designed to prohibit excessive rod removal were also employed.

Still, the final report on the investigation of the disaster by the Atomic Energy Commission placed responsibility on the shoulders of the three servicemen by concluding that the accident was due to "...manual withdrawal by one or more of the maintenance crew of the central control rod blade from the SL-1 core considerably beyond the limit specified in the maintenance procedure..."

---

their part in the rescue operations: Hazel M. Leisen, USAEC; William P. Gammill, USAEC; Edward J. Vallario, USAED; Captain. R. L. Morgan, U.S. Army; M. Sgt. Richard C. Lewis, U.S. Air Force; Paul R. Duckworth, Lawrence Radiation Laboratories; Sidney Cohen, Phillips Petroleum Company; Lovell J. Callister, Phillips Petroleum Company; Delos E. Richards, Phillips Petroleum Company; and William P. Rausch, Combustion Engineering Company.

REFERENCES AND NOTES

A.E.C. criticizes itself on safety (1961). *The New York Times,* June 11, 24.

A.E.C. opens inquiry into nuclear blast (1961). *The New York Times,* January 6, 8.

A.E.C reports status of Idaho accident investigation (1961). *Science,* January 27, 265.

And three were dead (1961). *Newsweek,* January 16, 74.

Armagnac, A. P. (1961). The atomic accident that couldn't happen. *Popular Science,* September, 52-55 and 214-215.

Atom aides scan effect of blast (1961). *The New York Times,* January 5, 19.

Atom victim recovered: third body taken from Idaho reactor building (1961). *The New York Times,* January 10, 20.

Atomic Energy Commission annual report to Congress for 1961 (1962). Washington, D.C.: U.S. Government Printing Office.

Atomic Energy Commission annual report to Congress for 1962 (1963). Washington, D.C.: U.S. Government Printing Office.

Fatal reactor blast in '61 laid to human failure (1962). *The New York Times,* September 25, 46.

Idaho atom blast laid to "runaway" (1961). *The New York Times,* January 20, 21.

Idaho nuclear peril limited (1961). *The New York Times,* January 9, 23.

IDO report on the nuclear incident at the SL-1 reactor (1961). U.S. Atomic Energy Commission, Idaho Operations Office, Technical Report IDO-19302.

Lewis, E. E. (1977). *Nuclear power reactor safety.* New York: John Wiley and Sons.

Operational accidents and radiation exposure experience within the United States Atomic Energy Commission (1968). Washington, D.C.: U.S. Government Printing Office.

Runaway reactor (1961). *Time,* January 13, 18-19.

Safety and the atom (1961). *The New York Times,* June 12, 28.

Seven atom specialists cited by hero fund (1962). *The New York Times,* October 12, 16.

SL-1 awards presentation ceremony - - program (1962). U.S. Atomic Energy Commission, Idaho Falls, Idaho, Idaho Falls Little Theatre, March 5, 8:00 P.M.

Thompson, T. J. (1964). Accidents and destructive tests. In *The technology of nuclear reactor safety* (Thompson, T. J. and Beckerley, J. G., eds.). Cambridge, Massachusetts: The M.I.T. Press.

Three killed by blast in atom reactor (1961). *The New York Times,* January 5, 1.

SILENT WARNING

Doctors Janabi and Amin barrelled down the two-lane road in their dusty white Peugeot toward the province of Kirkuk northeast of Baghdad. Amin was making good time. He had to if they were going to arrive at the clinic at a half-decent hour.

The scenery outside the car was barren and featureless. Janabi could not help but fill his mind with the unfolding events of the past few days and what might lie before them that evening. Never in his career as a physician had he faced anything so potentially serious, so unimaginably pervasive. If his suspicions were correct, history was being repeated. His call for action more than a decade before had gone unheeded. Hundreds - - perhaps thousands - - of Iraqis might have to pay with their lives for the mistakes of others. And it was all so easily preventable, so sickeningly unnecessary.

Dr. Janabi and his associates at the University of Baghdad's Department of Medicine learned of the situation only five days before when they received an urgent call from the physician at a medical clinic in the Kurdish province of Kirkuk. The rural doctor, learned Janabi, was swamped with dozens of patients from local farming communities. Nearly all had come in during the previous week. Most of the people were in very serious condition, some were critical. They were suffering from a wide range of neurological disorders with symptoms the young country doctor had never seen. Patients complained of sensory and motor losses, their hands and feet feeling numb and lifeless, their motions clumsy and stumbling. Those who could still speak were strangely irritable and anxious. Others could not walk, and a few were losing their sight. Some had even lost their hearing. The worst cases were comatose and on the verge of death. The doctor believed he was dealing with an epidemic of encephalitis, but Dr. Janabi suspected otherwise. If he were correct, and he sincerely hoped that he was not, they were facing possibly the largest mass poisoning in history. It was certainly unintentional, yet all so avoidable.

The call from the country doctor five days before was just the beginning. The University had fielded hundreds of telephone calls from nearly every province in Iraq in the days following the young man's call for assistance. Four hundred new cases were appearing at hospitals around the country each day; they were principally from the central areas of Babylon and Qadissiya and the northern Kurdish provinces. The common denominator, they had determined, was that all of these victims were farmers or members of large farming families. And, like other Iraqis, they ate a great deal of bread.

There would be only one thing directly responsible for the disease, reasoned Janabi. But as they continued the long drive to Kirkuk, he mulled over the odd set of circumstances that had worked together to bring about this catastrophe. First, as strange as it seemed, one had to consider the harsh weather during the past two years.

It had been brutally dry for the last couple of summers and the farming areas, especially those up north, were affected most severely. Supplies of grain, particularly wheat, ran dangerously low throughout the country. Some rural farmers, he was told, exhausted their supplies of wheat and dipped into their reserves of valuable seed grain to feed their families. They addressed their immediate problems, but only at the risk of not having enough seed to plant the next crop. It had been a serious situation, but fortunately (or unfortunately, as time would tell), the Ministry of Health had stepped in to tackle the situation. And that, most certainly, was the second important element of the story.

Janabi knew now that the Iraqi Government and Cargill Corporation of Minneapolis, Minnesota struck a deal over a year before, in 1970, for a large purchase of seed grain. The shipment of 73,201 metric tons of wheat and 22,262 tons of barley was delivered to Iraq from North America and a few other sources during the fall of 1971. It arrived at the port of Basra in two large shipments between September 16 and October 15, 1971 and was then trucked to regional distribution centers throughout the country, but mostly north to the Kurdish provinces of Ninevah, Arbil, and Kirkuk where it was needed most urgently. The shipments would make it possible for rural farmers to grow their own crops of wheat and barley and to feed their families and animals with the grain they grew. It was enough seed grain, Janabi had recently learned, to meet the planting requirements of all of Iraq for the year, but especially for the crop of winter wheat that needed to be sown in October, November, and early December, before the start of the winter rains. As these things often go, however, the grain was not distributed as quickly as planned, and by the time it made it to the farmers in November and December many had completed planting with their own supplies of seed. Some of the newly imported seed was planted, but a huge portion remained unused.

Dr. Amin continued driving at a blistering pace toward the rolling mountains ahead. The sun hung just on the horizon off to their left. It would soon be dark, and traffic on the rural highway was beginning to thin. Janabi guessed that they would be the only ones still on the road when they arrived at the clinic late that night. It would be at least another three or four hours. He appreciated not having to drive; it was the first time in days that he had had the chance to sort out his thoughts.

Yesterday he heard that the Iraqi Ministry of Health might have specified that the shipment of grain be treated with alkylmercury fungicide when they made the deal with Cargill the year before. It was one of a number of substances commonly applied to seed grain to inhibit spoilage. As with other seed dressings, however, treatment with alkymercury rendered the grain unfit for consumption; and, accordingly, it had been dyed red as a warning of its toxicity. As an extra precaution, a large square tag was stitched into the seam of each large sack of grain. The label contained a strong warning that the grain could not be consumed or milled into flour, but it was printed in English, not Arabic. A large image of a skull and crossbones sat above the text to emphasize the point.[1]

[1]Cargill, like other grain dealers in the United States, could not sell

But the Kurdish farmers, not surprisingly, did not read English. Most, in fact, did not read Arabic. And the large skull and crossbones symbol on each tag was nothing more than a peculiar piece of art work. Unlike Westerners, the symbol did not connote death or danger to the Kurdish peasants. It most definitely did not warn them that the grain was exceedingly poisonous.

The fourth and certainly most tragic part of the story, knew Janabi, was that thousands upon thousands of Iraqis, mostly Kurds, had possibly eaten the treated grain in one form or another. He had been informed that peasant women, while their husbands were out working in the fields, washed the grain in large bowls to remove the red dye. Some farmers were told at the distribution centers that the grain was chemically treated, and the peasants reasoned that washing the red color from the seed removed the chemical as well. But the alkylmercury was not physically bound with the dye. It remained behind on the small seeds after the red colorant was rinsed away. Neither the dye nor the alkylmercury had an intolerable or even noticeable

---

this type of dressed seed in the U.S. as a result of recent action by federal courts and the newly enacted Federal Environmental Pesticide Control Act of 1971 prohibiting interstate shipment of seed treated with alkylmercury fungicide. American officials, especially their counterparts in Sweden, had become aware of the hazards of the alkylmercurial fungicide seed dressing after accidental poisonings of farmers in Sweden, Guatemala, Russia, and Pakistan, and in Iraq on two separate occasions during the previous 20 years. All of these cases involved rural farmers either feeding the seed grain to animals, which were later consumed, or grinding the seed into wheat for bread. The highly publicized environmental and health catastrophe in Minimata Bay and Niigata in Japan, although unrelated to treated seed, had also added to the concern about the growing presence of methylmercury in the environment.

taste or odor that might dissuade a person or animal from eating it.

Many farmers, some before and some after rinsing the dye off the seed, fed the grain to chickens and goats to test its safety. But organic mercury causes no visible symptoms until it accumulates and reaches a critical level in the body. The farmers, Janabi reasoned, had continued to feed their animals for weeks on end with no ill effects, unaware that the effects were long delayed.

The grain, he was nearly certain, was then milled into flour on the hundreds of primitive grinding stones scattered throughout the farming communities. It was then formed into thousands upon thousands of pancake-like loaves and baked in small outdoor ovens. Each loaf, as best as the physicians at the University could determine, probably contained about 1.4 mg of methylmercury (a type of alkylmercury), and each person ate about three loaves each day. It was late December and early January when the first symptoms surfaced for most people who had been eating the bread. This was the point at which the consumed methylmercury reached a critical threshold, or "body burden." There had been no major symptoms before this time because most people had not yet reached the threshold concentration. But now the story was different. Victims were flooding into the rural clinics such as the one in Kirkuk.

He knew that organic mercury left the body very slowly. Its half-life was more than three months. Once consumed, it continued to work its damage on the nervous system for months and even years, if the patient survived that long. And the damage was permanent and grotesque.

The most tragic thing of all was that little of what Dr. Janabi had learned during the past few days came as a surprise. Ironically, it had all happened before. Just ten years prior, he and his colleagues published warnings for the world's medical community about the dangers of distributing treated seed grain to peasant farmers without proper safeguards, warnings, and education. Some physicians and scientists might have understood the implications of the advice, but it was painfully obvious that the traders and bureaucrats failed to heed their warnings.

It was close to midnight when they arrived at the clinic in Kirkuk. The young doctor, obviously anticipating their arrival, met them outside as they walked toward the building from the car. Janabi asked to see a few of the patients prior to discussing the situation or resting from the long trip. The main ward was filled to capacity. Many of the patients were young women, all were lying on beds and cots. He walked quietly to one and gently opened her contorted and now useless hands. Faded traces of red dye lined the creases of her palms.

EPILOGUE

Between January and the end of August, 1972, 6530 admissions and 459 deaths from methylmercury poisoning were recorded in hospitals in Iraq. Based on expert opinion and general observations by tourists, however, it was estimated that as many as 60,000 peasants ingested sufficient methylmercury to have suffered neurological damage. There may have been as many as ten deaths for every official death recorded in hospitals.

With overwhelming evidence of the catastrophe underway, the government placed a ban on the sale of meat, recalled all of the seed grain distributed that fall, and mandated the death penalty for anyone caught selling the grain. They also instituted a news blackout to the rest of the world to avoid embarrassment. Officials eventually confiscated 5000 metric tons, but over 90 percent of the original shipment of grain remained unaccounted for. Much of it, they believed, was dumped into streams and rivers by peasants fearful of being arrested for having it in their possession.

New admissions to hospitals began to taper off by the beginning of March and it appeared that the cause of the catastrophe had been identified and dealt with. Thousands of individuals remained hospitalized for the remainder of the winter and spring, however, and the deaths continued to mount as the severely stricken succumbed to the irreversible damage done by the poison.

REFERENCES AND NOTES

Bakir, F., Damluji, S. F., Amin-Zaki, M. L., Murtadha, M., Khalidi, A., Al-Rawi, N. Y., Tikriti, S., Dhahir, H. I., Clarkson, T. W., Smith, J.C., and Doherty, R. A. (1973). Methylmercury poisoning in Iraq. *Science*, 181, 230.

Damluji, S. and Tikriti, S (1972). Mercury poisoning from wheat. *British Medical Journal,* 1, 804.

Dillman, T., and Behr, P. (1972). Mercury poisoning in Iraq termed a disaster. *The Lansing State Journal*, Lansing, Michigan, March 25.

D'Itri, P. A. and D'Itri, F. M. (1977). *Mercury contamination: a human tragedy*. New York: John Wiley & Sons.

Iraqi Poisonous seed might have originated in Canada: Official (1972). *The Toronto Globe and Mail*, March 10.

Jalili, M. A. and Abbasi, A. H. (1961). Poisoning by ethylmercury toluene sulphonanilide. *British Journal of Industrial Medicine,* 18, 303-308.

Nelson, G. (1972). *Statement of U.S. Senator Gaylord Nelson on H. R. 10729, the Federal Environmental Pesticide Control Act of 1971, and amendments, before the Senate Commerce Subcommittee on the Environment, Senator Philip A. Hart, Chairman,* June 15.

The story line is based on published accounts and is believed to represent the experiences of the physicians and researchers at the University of Baghdad who responded to this catastrophe. The names of the characters in this story are fictitious.

ZZZs IN ZEEBRUGGE

Assistant Boatswain Mark Stanley finished tying down the last of the large truck trailers on G deck and returned to his cabin for a short break from work before the scheduled departure for Dover on the other side of the English Channel. His 24-hour work shift made the day long and dreary, and it was refreshing to sit down alone for a few minutes and get warm. He drank a cup of hot tea, closed the porthole to shut out the damp Belgian air, and sat back down on his bunk to read for a few minutes until the call to report to harbor stations. His job was to return to G deck and close the large bow doors of the ship as they got underway - - if it had not been done already by someone else, as was often the case. But the warm tea in his stomach and the quiet isolation of the room began to work its spell, and seaman Stanley, outfitted in his grimy orange coveralls, slouched against the wall next to his bunk and fell fast asleep aboard the *Herald of Free Enterprise*.

David Lewry, Captain of the ship, stood in the wheelhouse a quarter of an hour later at 7:00 p.m. and prepared to guide the big ferry out of Zeebrugge and across the channel for the nightly run on March 6, 1987. Thc sun had set, and the evening was quiet and cold. The green phosphors of the radar screen cast a soothing glow over the darkened room. Lewry looked out over the water; nothing was the least bit unusual and it should be an uneventful trip.

The call for all hands to report to harbor stations was announced over the loudspeaker a few minutes before departure. A message was received from Chief Officer Leslie Sable that the shore ramp was being lifted. The radio crackled with a message from Zeebrugge port control stating that the harbor channel was now clear of traffic and that the *Herald* could be on her way. Captain Lewry issued the remaining orders to the chief, second officers, and two quartermasters there with him on the bridge. They were the same orders that he had given hundreds of times before, and they were acknowledged by the men on the bridge in muffled obedient voices.

Captain Lewry walked outside to the wing of the bridge and watched the stern of the ship as she backed out of her berth. He did not turn around to look at the dock and the front of the ship; but, if he had, he would not have seen that the bow doors were open. They were blocked from his view. And, if he had walked back inside to the control console, he would not have seen any indicators telling him that the doors were open because there were no door-position displays. So the *Herald* pulled back from the ferry terminal with everything apparently in order, pointed her bow toward England and steamed slowly through the harbor at Zeebrugge and into the cold water of the North Sea.1

It was a Friday night, and the crowd aboard was a collection of tourists, business people, truck drivers, and servicemen. Nearly all of them were British, many returning to the U.K. after a holiday on the continent or, in the case of the soldiers, going home for a weekend of leave from their station near the Rhine. All told, there were 460 passengers, 80 crew members, 81 cars, and 47 trucks. It was not the full capacity of the ship, but definitely a heavy load. By now the passengers were spread throughout the decks. Many lined up at the duty-free shop, and even more were finding a place to sit and have a meal in one of the two large cafeterias. Some, like seaman Stanley, had already dozed off to sleep for the four-and-a-half hour trip to Dover.

As the *Herald* maneuvered through the corridors of the harbor and past the sea wall, her two forward pumps were busy

[1] Although Lewry assumed that the bow doors were closed or, at the very least, that the crewmen responsible for closing them would do their job, sailing through the harbor with the doors open was not all that unusual for the ferries of the Townsend Thoresen line. Like other "drive on - drive off" ferries, the two vehicle decks on the *Herald* filled with vehicle exhaust during loading, and it was necessary to clear the air in the expansive, flat holds. The 12-ton bow and stern doors were frequently left open as they steered through the harbor to blow the foul air out of the ship. Some captains of the fleet viewed this as a potential hazard and had voiced their concerns with management. Five ferries were known to have sailed with their bow doors open since 1983, and another, the *Pride of Free Enterprise,* was known to have sailed with both the bow and stern doors opened to the sea. Her captain had taken the bold step of writing to management, requesting that indicator lights be installed on the bridge so that the position of the doors would be known at the helm. One manager of the line, responding to this captain's suggestion, responded by saying "Do they need an indicator to tell them whether the deck storekeeper is awake and sober?"

pumping the water from the large bow ballast tanks and returning it to the ocean. The same forward tanks had been filled with water when she approached the dock in Zeebrugge some hours before. This brought the forward loading ramp of the ship down about three feet to the same level as the dock. But now that the ship was underway, it was necessary to activate the pumps to remove the water and raise the bow of the ship for the voyage across the channel. She was still nose-down, however, because the pumps were of limited capacity and took nearly two hours to expel the water and raise the bow to the normal level.[2]

Captain Lewry dismissed two officers on the bridge so that they could take their breaks for dinner just as the *Herald of Free Enterprise* left the harbor at Zeebrugge. He then ordered the helmsman to pick up speed; they would slowly work their way up to the top rate of 22 knots. But with each small increment in speed, the bow wave rose closer and closer to the open doorway. Finally, it bubbled up underneath the "spade," an extension or tongue of the forward vehicle deck used during off-loading. When the *Herald* was just one mile outside the breakwater, she reached a speed of 15 knots, and the first gallons of water crested

[2] Six months before, Captain Lewry wrote a memo to other captains and engineers who had served on the *Herald* to open a dialogue on a problem with the forward ballast pumps. Lewry, as well as other officers, noted that the pumps on their German-built ship were of very limited capacity and often took two hours to clear the water from the forward ballast tanks. As a consequence, the ship rode in a nose down position for a good part of the voyage, a situation that was especially disconcerting in the heavy swells of the North Sea. Lewry noted in his letter that sometimes "the bow wave is well up the bow doors." Although there was talk of retrofitting the ship with higher-capacity pumps, neither the British Marine Department nor the management at Townsend Thorensen Company had acted on the matter.

the spade and rolled freely into the car hold. The flow of sea water transformed into a solid and thick wall within ten seconds. It rolled up the bow and directly through the open doorway, leveled out as it entered the ship, and spread across the flat forward deck at the rate of 200 tons a minute.

Once onto the floor of G deck, the water ran across the open expanse of the football-field size room. Unhindered by vertical bulkheads, it gushed across the floor and under the vehicles, some of it running back 433 feet toward the stern and down the stairways into the H deck. The entire surface of G deck was soon deep in water, and, like the fluid on a large shallow tray being carried across a kitchen, the tons of water on the deck could shift and gush to one side at any moment, instantly toppling the huge ship. The inevitable happened within a minute, and the massive weight began to shift to the port side, just slightly in the first seconds and then in a torrent.

Captain Lewry felt that something was not quite right about how the ship was behaving, but he could not tell what it was. He looked over at the helmsman, Quartermaster John Hobbs. Hobbs was spinning the wheel rapidly to the left! The helmsman had a frozen, horrified look on his face.

Hobbs, without looking at Lewry, cried out, "I've got the helm on port and she's going to starboard!" Unknown to Lewry and to Hobbs, the mass of water had slid across the lower flat vehicle deck, drastically shifting the center of gravity of the ship and twisting her course suddenly to the right. In response, Hobbs had countersteered to the left, but the ship was continuing to steer uncontrolled to starboard.

"What on earth is going on?" yelled Lewry. Abruptly, the floor, the control consoles, and indeed the entire bridge began to slide down and to the left. The right side of the bridge rose up, and loose items in the room slid across the 40-foot width of the bridge, accelerating as the listing continued.

Lewry was terrified, yet he had only one thought: stop the ship or, at the very least, do something - - anything - - that might stop them from going all the way over. He jumped forward to the propulsion controls on the starboard console on the far right side of the bridge. The floor now tilted some 20 degrees to the left. He pulled the handles to reverse the pitch on the three main propellers at the stern. But Lewry could only hold on to the handles as the floor continued to sink to the left.

Incredibly, it seemed to stop as quickly as it had begun, and Lewry sensed that the ship had halted its list, although they were now tilted at nearly 30 degrees. They might get out of this after all. For three or four critical seconds the ship hung in the balance, not indicating which way she was going to go - - back to where she was just a few moments before or the unthinkable alternative. But then the mass of sea water in the ship made one inevitable and fateful shift, and the 8,000 ton *Herald of Free Enterprise* and the hundreds of tons of sea water sloshing about in her lower decks rolled just past the state of precarious equilibrium. Lewry knew instantly that she was going over, and there was nothing anyone could do about it!

The floor tilted instantly, the port side dropping away beneath his feet. He fell freely through the darkened room along with the other officers and tons of equipment. For that dreadful second Lewry could think of nothing but keeping his feet and legs pointed downward as he descended. He hit the port wall of the bridge 40 feet below and felt something hard and sharp pierce the left side of his chest. Unable to move and incapable of taking the pain, Captain Lewry lost consciousness just as the dark icy water of the North Sea exploded in through the broken windows of the bridge.

Seaman Stanley, still asleep on F deck, was thrown off of his bunk as the ship rolled over on its side. He realized immediately that the *Herald* was at sea and, based on the angle of his cabin, that she had capsized. He knew also that he had slept through the call to harbor stations. After getting to his door and into the hallway, he made his way along the corridors to the starboard side of the ship. By now the *Herald* had settled on her port side in about 30 feet of water with about half of the ship below the surface. There were screams from behind every window and port hole along the starboard walls now exposed to the dark sky. Stanley grabbed an ax from a lifeboat and began to break the glass to the rooms beneath, only to find that the passengers were floating in the freezing water 30 feet below. Many were injured or unconscious, and most had nothing to help keep them afloat. Assorted ropes and ladders were located, and Stanley and other able-bodied people who had made their way to the starboard side of the ship began to rescue passengers trapped in the water-filled rooms below. They were soon aided by Zeebrugge emergency crews and Belgian military personnel. Among those rescued was Captain Lewry. Less fortunate were the 188 passengers and members of the crew who perished.

EPILOGUE

Later that year the British Court of Inquiry found that the accident was attributable to the actions of Assistant Boatswain Mark Stanley, who failed to close the bow doors, to Chief Officer Leslie Sabel, who failed to verify that the doors were closed, and to Captain David Lewry, who should never have left port without being certain that the bow doors were closed. However, the Board's most potent remarks were directed at the company, Townsend Thoresen, which was guilty of running a generally "sloppy" operation.

By early April, the *Herald of Free Enterprise* had been refloated, and the last of the bodies inside were recovered by divers. She was then renamed and sold to a Taiwanese shipyard for scrap. In the interests of presenting a fresh face to the public, Townsend Thoresen changed the colors on the other ships of the fleet from the red, white, black, and orange of the *Herald* to a soothing light blue. And after the government's inquiry that summer, all British channel ferries were required to install indicator lights and video monitors so that officers on the bridge could determine the position of the cargo doors before leaving port. But by October the memory of the greatest British peacetime disaster at sea since the sinking of the *Titanic* had faded: there were five reports of channel ferries getting underway with their doors open, presumably to disperse exhaust fumes.

REFERENCES AND NOTES

A fading tragedy (1987). *MacLeans,* November 9, 8.

Along Flanders coast, prayers for victims (1987). *The New York Times,* March 9, 3.

As ferry survivors weep, 408 are safe in Belgium (1987). *The New York Times*, March 8, 1.

Auto ferries vulnerable to flooding (1987). *The New York Times*, March 7, 6.

Britain begins inquiry into accident; roll-on, roll-off design questioned (1987). *The New York Times,* March 8, 16.

Britain opens ferry inquiry (1987). *The New York Times,* April 28, 6.

Buck, L. (1989). Human error at sea. *Human Factors Society Bulletin*, September, 32 (9), 12.

Damage to ferry debated in Britain (1987). *The New York Times,* March 14, 3.

Elliott, L. (1988). Night of anguish, night of courage. *Readers Digest*, March, 132, 115 (28 pages).

Ferry investigation focuses on the open loading doors (1987). *The New York Times,* March 9, 3.

Ferry with over 500 people capsizes near Belgian port; 350 are safe, many trapped (1987). *The New York Times,* March 7, 1.

Heroism - and horror (1987). *MacLeans,* March 23, 54-55.

Many are missing after ferry capsizes (1987). *The New York Times,* March 7, 6.

Safety's price (1987). *The New York Times,* March 12, 30.

Speedy rescue effort saved many aboard capsized ferry (1987). *The New York Times,* March 9, 3.

Sunken ferry righted off Belgium (1987). *The New York Times,* April 8, 12.

Survivors, sobbing with relief, reach British Airport (1987). *The New York Times*, March 8, 16.

Tightened rules seen for ferries (1987). *The New York Times*, March 9, 3.

Tragedy in the harbor (1987). *MacLeans*, March 16, 26-27.

U.S ferry standards called tougher (1987). *The New York Times*, March 8, 18.

408 saved in capsizing of British ferry (1987). *The New York Times*, March 8, 16.

DOUBLE VISION

Dr. Mark Anderson turned away from the x-ray film at his workstation and glanced back at the large group of people entering the room from the hallway outside. The one in front was Dr. Barbara Zador, Staff Radiologist at the Regional Medical Center. Right behind her were two local pediatricians, then a dozen more people - - students, it looked like. There could not have been twelve more straight-laced and well-scrubbed young physicians in all of Southern California. Each of the women had her hair perfectly done, and even the men wore the whitest and most thoroughly-pressed coats Mark had ever seen. Dr. Zador directed the group over to the viewing station next to Mark. Yes, these were definitely the new interns from pediatrics. They were going to be given a tutorial about a case.

Mark turned back around and directed his attention once again to the x-ray on his "board," one of ten workstations in the radiology reading room there at the center. Each workstation was dedicated to a specific function, and that particular morning he was viewing about 100 x-rays on the "chest" board. Appropriately enough, the image in front of him was that of a chest, and, like most of the thousands upon thousands of films he had viewed during his five years of residency in radiology, there appeared to be nothing at all noteworthy about this particular image. It was completely normal. He wrote a brief note with his pen and pressed the button to advance to the next image on the roll.

To Mark's immediate left was the "bone" board, used almost exclusively for shots of hands, feet, legs, and arms. And behind him on the other side of the room were the "neuro" board and the "GI" or gastro-intestinal board. Some boards, like his "chest" board, handled x-rays sandwiched between sheets of clear plastic in a roll. He could view up to 100 films in a sitting by pressing a button and letting the machine advance the pictures. Other workstations, like the pediatrics station to his right, were simple back-lit light boxes with clamps for holding standard x-ray sheets.

Mark and the group from pediatrics were the only physicians in the radiology room that morning, so it was difficult not to overhear the lecture on the case. Dr. Zador placed two films up on the light panel and began a review of the patient's history. They were images of the thigh of an eight-year old boy. One view was an AP or frontal view of the upper leg; the other view was from the side. As in nearly every case, two views were required in order to comprehend the location of points in three dimensions. Two views were also necessary to determine the volume or depth of findings seen on the films.

It was a particularly intriguing case, and the interns huddled around Dr. Zador and the x-rays. Each young doctor wanted a good look at the images. A number of them stood with their backs just inches to the right of Mark's chair. He could not do a thorough job reading the next chest x-ray on his own workstation with all of the distractions, so he quietly pushed back and stood up at the back of the group. Anyway, this wasn't a bad time to stretch his legs - - and this sounded like it could be interesting.

The two pictures were clipped to the light board about a foot overhead, so Mark had a reasonably good view of the large transparent films even though he was standing behind everyone else. This was Dr. Zador's show, and his intentions were to listen in and learn something. She began her description.

"Here we have a case of calcification of the soft tissue of the thigh."

As she talked she reached up and traced the outline of an elongated image with her finger. The irregular form meandered down the middle of the front-view x-ray, generally surrounding the femur, or thighbone, in the middle of the boy's leg. The side view provided a clear shot of the form as well and showed the calcification to be roughly as deep as it was wide.

"As you can see on both the AP and lateral views, we have curvilinear calcifications in the major muscle groups of the thigh. No bone destruction or periosteal reaction is evident."

This was indeed an unusual case. He had never before seen anything quite like it.

She continued with the analysis. Dr. Miller, one of the two non-resident pediatricians with her, brought the case to her attention the previous week. The patient, actually the boy and his mother, came to him complaining of mild hip pain. X-rays were taken and the condition identified.

"This condition is due to one of three diseases, but we cannot be absolutely certain which it is until further tests have been completed." She paused for dramatic effect and then continued with her lecture. "First, it is possible that this is a case of dermatomyositis with calcinosis universalis. As you all know, it is a connective tissue disease like scleroderma. We can see the thin calcific plaques throughout the soft tissue of the leg," she said, pointing to the unusual swirls within the leg on the x-rays.

Mark's attention focused on the two images during Dr. Zador's review. Her mention of dermatomyositis with calcinosis universalis raised his curiosity about the case another notch. He had never seen an instance of dermatomyositis with calcinosis universalis in a boy this young. Strangely, there were no other symptoms. Nothing on the skin as would be expected in dermatomyositis.

Dr. Zador resumed her lecture. "The second and more likely diagnosis is that this is due to Loiasis, a parasitic infection attributable to the filarial nematode Loa loa. The swirls we see here are the calcified remnants of a worm in the soft tissue."

Loa loa? Mark was very familiar with this ghastly affliction. He had always had an interest - - more of a gruesome curiosity - - in unusual diseases and remembered studying this one. While a student he had even thought seriously about a career in parasitology but decided that there was only a limited market for such knowledge, at least in the U.S. Radiology seemed like a surer bet.

Dr. Zador's summary of Loiasis was characteristically accurate, thought Mark. But it was the part that she didn't mention that was so relevant to this particular case. The carrier, the day-biting mango fly, deposits infective Loa loa on the skin of an unsuspecting person. The minute organism then penetrates a hair follicle, develops into an adult in the subcutaneous tissues, and wanders throughout the soft tissues of the body. Symptoms include fugitive swelling over bones, mainly the wrist. The adult worm occasionally passes in front of the person's eye, thus the basis for the disease's common name, African eye worm. But Dr. Zador had not mentioned that the Loa loa worm is found only in the rain forests of tropical Africa.

It seemed highly improbable that a Caucasian boy from Southern California would have a case of Loiasis, especially a boy who had never traveled outside the country. And the boy had none of the other symptoms of the disease. The swirls on the film were certainly peculiar, and Dr. Zador's rundown on Loa loa and its symptoms was accurate, but this could not possibly explain what was on the x-rays. She was analyzing a portion of the facts, only the facts that matched a limited set of the symptoms of the disease. And the symptoms, Mark had a hunch, were suspect.

It was common for radiologists to review and discuss one another's work there in the film viewing room, but in this particular instance an uninvited comment about the diagnosis would have been rude. Dr. Zador was Chief of Pediatric Radiology and, in effect, one of Mark's superiors. She had been doing this for many more years than he. She would not appreciate even the slightest challenge to her diagnosis, especially in front of a dozen new pediatric interns. He continued to examine the films, but remained quiet.

"Our third hypothesis," Dr. Zador continued, "is that this is a case of dracunculiasis medinesis. It, too, is a parasitic infection that can result in swirling calcified structures in deep soft tissue, usually in the lower limbs."

Mark's forbearance had worn thin. Dr. Zador's diagnosis was plausible, given her selective treatment of the details of the case, but not at all probable. Dracunculiasis, commonly known as Guinea worm infestation, is also a disease of the tropics. Its root cause is the consumption of drinking water contaminated with infected cyclops. Released larvae penetrate the wall of the intestine, he knew, then mature, mate, and travel to subcutaneous connective tissue, usually in the legs and feet. The worm grows to more than a meter in length. It then protrudes from the skin and excretes larvae. A dead worm sometimes calcifies, leaving swirling patterns detectable in x-rays. It was this last fact on which she had obviously fixated. But Mark knew her conclusion was not tenable in view of all of the information at hand.

Mark had now examined the images for about five minutes, looking first at one, then the other, and then back to the first. He studied the volume of the structure, based on the front and then the side view, and the path of each of the major lines and curves which were so prominent throughout the edges of the strange form. The structure had a definite organic appearance with numerous curves and linear densities, yet at the same time it was so... clean.

Yes, that was it. The structure in each of the two films was subtle and natural-looking, but the outlines and densities within the images were unusually well-defined, almost segmented. They looked almost like concentric drops of paint that someone had tried to wipe off. Yet they were on both films and therefore three-dimensional.

Without saying anything, Mark walked the few steps around the group to the side of the light board and moved the two images closer together. He then stood back a few feet to the side, but not as far back as his previous position behind the pediatric interns.

The answer was there in the films all this time, but it came to him abruptly and unexpectedly. The two faint and mysterious images, one on the front film and the other on the side or lateral film, were of the same thing. That's what was so odd about these films! And the images were not just of the same thing, they were identical. This would be possible only if the thing responsible for the image was perfectly symmetrical, such that the same image would be present in front and side views. But that was nearly inconceivable. They were more likely identical pictures of the same flat two-dimensional object. This would be possible only if the mysterious object was actually outside the boy's leg and if the leg had been placed under or over the object when the x-ray was taken. It was the only explanation.

All the while Dr. Zador continued her review of the case,

stating that four experts had examined the films and that they all agreed with the diagnoses, with general consensus on the last one.

"We are reasonably certain that this is a case of dracunculiasis medinesis. We will be performing a surgical biopsy early this afternoon."

His pulse quickened upon learning of the impending biopsy. Surgeons would remove muscle from the leg and examine it for signs of disease. This had gone too far, and he decided, politics and sensitivities aside, that he could not let this one go by. They would find the boy's leg to be disease-free after the biopsy, but it would be inexcusable if he said nothing and allowed the patient to undergo the unneeded procedure. All invasive operations involved some level of risk. Mark had to stop this from going any further and do so in such a way as to preserve his professional relationship with Dr. Zador.

He was determined to make his point, but he did so hesitantly. "Uh, Barbara...."

She stopped before beginning her next sentence. Dr. Zador had obviously been aware of his presence throughout the lecture, and she responded rather curtly. "Yes, Mark?"

"Please pardon me for interrupting, but I don't think these calcifications are real. This is an artifact."

"What do you mean?"

"This is an artifact, an anomaly. This is not soft tissue calcification. These two images are the same, and I am certain that this is not an image of something in the leg."

She was not happy about being interrupted, and visibly irked that he had questioned the diagnosis, especially in front of her captive and impressionable audience.

"We have had these films read by four experts, and we are all in agreement. I don't think there is any question about what we are dealing with. There are some questions as to the cause, but.."

Just then she was interrupted by one of the young interns. Mark had seen him nod his head in agreement a moment before when he began to talk.

"No, he's right. Look, this pattern of swirls is the same on both films," he said, as he stepped forward and touched the two images simultaneously with his left and right hands.

Mark sensed that others in the group had seen the same thing. A few of the pediatric interns began whispering to one another.

No longer the lone voice in the crowd, Mark stepped forward to the light board, slid his fingers under the film on the right, and moved it over to the left directly on top of the other large film. He adjusted it slowly, holding it between his thumb and fingers, until the unique swirling image at the lower portion of the x-ray matched up perfectly with the same image on the x-ray underneath. Sure enough, they were the same. Mark looked back at the interns, some of whom were grinning slightly, and then back over to Dr. Zador. She was not smiling.

Although Dr. Zador and the two attending pediatricians did not acquiesce to Mark Anderson's argument there in the radiology viewing room, they did agree on the appropriateness of canceling the biopsy scheduled for that afternoon. The x-ray would be redone - - but this time on a different machine.

When the boy's leg was x-rayed again, the calcified remnants of the Guinea worm had mysteriously disappeared. A follow-up investigation of the original x-ray machine found that a small amount of x-ray contrast solution, a special dye used in some procedures, had been spilled on the film guiding mechanism underneath the table some time in the past. An interesting pattern of lines and rings was left as the solution dried, much like the rings on a bath tub once the dirty water has drained away. It was also smeared, probably by internal moving parts or perhaps by someone trying to clean up the small spill. The image had been superimposed onto the image of the leg in exactly the same location in both the front and side view films. There was nothing significantly wrong with the boy's leg. Drs. Mark Anderson and Barbara Zador each had a renewed appreciation of the importance of the second view.

REFERENCES AND NOTES

The names of the characters in this story are fictitious. Any similarities to the actual names of these individuals is purely coincidental.

THAT NEWFANGLED TECHNOLOGY

It all began in the afternoon, September 7, 1923. Lieutenant Commander Donald T. Hunter, Captain of the *USS Delphy*, stood at the window of the quiet wardroom aboard the *USS Melville*, captivated by the magnificent panorama across San Francisco Bay. The golden light seen only along the California coast late in the day at the end of summer washed over the harbor and surrounding hills. Moored nearby were the 14 stately gray ships of Destroyer Squadron 11, anchored peacefully on the still water of the harbor and illuminated by the low yellow sunlight. A flotilla of small wooden gigs rowed out from the destroyers to deliver the 40 command and engineering officers of Destroyer Squadron 11 to the *Melville* for the meeting concerning tomorrow's training exercise. Synchronous strokes of the oars propelled each boat toward Captain Hunter with speed and grace. He had seen this beautiful spectacle of form and color a few times before during visits to the Bay, and he knew that it would last only a few minutes more. The light would soon begin to dim ever so slightly, just enough that the saturated colors of the mountains, sea, and ships would fade, and the scene would begin its slow advance toward nightfall.

"Exquisite view," commented Captain Edward Howe Watson, Commodore of the Destroyer Squadron. He was standing just behind Captain Hunter's right shoulder.

Hunter, a bit surprised that Commodore Watson had managed to get so close without him noticing, didn't miss a step in responding. "It certainly is. If you must be in port, this is definitely the place to be. It's difficult to imagine a more serene spot." The gentlemanly Commodore smiled and nodded his head in agreement.

The first of the gigs had reached the starboard side of the *Melville* below, and both men turned their attention to the boarding operations. Four tanned, white-suited officers jumped expertly from the first rowboat, climbed up the netting on the side of the big ship, and raced smartly up the gangway and to the wardroom. Hunter sensed that each man knew he was being watched from above as he boarded the *Melville*. Indeed, Hunter and Watson scrutinized the operation from the wardroom, and Rear Admiral S.E.W. Kittelle, Commander Destroyer Squadrons, Battle Fleet and on whose ship they were meeting, observed from the deck above.

Within a few minutes the last of the 40 Navy officers of Destroyer Squadron 11 entered the wardroom and took their seats. Captain Hunter turned to Commander Watson, seated next to him at the head of the big teak conference table, and nodded his head in agreement, silently signaling that everyone was present and that they were ready to proceed. A bell rang out six times, signaling the hour, and Captain Hunter quickly glanced around the table to see that all of the other officers were ready.

Commodore Watson called the meeting to order and got straight to business. Hunter knew all of the details, as did most of the other officers, but he sat attentively as Commodore

Watson begin to speak. "On the command of Admiral Kittelle, who, incidentally, has been kind enough to let us use his wardroom for our meeting, we shall continue the Fleet Exercise tomorrow with a high-speed run down to San Diego. We will depart San Francisco midmorning and assume a speed of 20 knots for the duration of our cruise. An hour or two after sunset we shall make a sharp left turn after passing Point Conception and sail eastward along the coast through the Santa Barbara Channel between the east-west coastline and the Channel Islands about 25 miles offshore. A few hours later we shall bear southward once again and follow the coastline down to San Diego. I will be sailing aboard the *Delphy* with Captain Hunter."

Commodore Watson filled in the details of the engineering run for the 40 officers during the remainder of the conference. In all, 36 ships of the U.S. Navy would set out for southern ports the next day. Some, like the *Melville,* would leave well before dawn due to limited speed or engineering problems. Other ships, like those of Destroyer Squadron 12, would take a more westerly route well off the coast. If all worked out according to the Commodore's plan, the 14 "four stackers" of Squadron 11 would be among the last to leave but the first to arrive in San Diego, 427 miles to the south, a day and a half away on the morning of the 9th. There they would greet the other ships of the Battle Fleet as they reached their home port.

Hunter sat proudly as the Commodore went on to explain that the ships of Destroyer Squadron 11 would steam southward with the *USS Delphy*, the Squadron Flagship, at the lead. She would be commanded by Captain Hunter. Commodore Watson would sail aboard the *Delphy*. They would be simulating war conditions and requirements, so the ships of the squadron were to follow in a line and maintain a distance of 147 yards between the bow and the stern of the ship in front. Only the low-power turbines would be used in order to conserve fuel, but they could still move the ships at the quick pace of 20 knots. Radio communications were to be kept at a minimum, and all ships but the Flagship were forbidden to request radio bearings from shore stations. The *Delphy,* in the lead, would be taking care of all navigation duties. The other destroyers were to hold to the course of the ship in front and maintain the assigned following distance. The cruising order was the *USS Delphy, USS S.P. Lee, USS Young, USS Woodbury, USS Nicholas, USS Farragut, USS Fuller, USS Percival, USS Somers, USS Chauncey, USS Kennedy, USS Paul Hamilton, USS Stoddert,* and *USS Thompson.*

Everyone in the room knew that all of the destroyers in the squadron were state-of-the-art and surprisingly capable. Captain Hunter's ship, the *Delphy,* was a typical ship of the class and had an overall length of 314 feet and a beam of only 32 feet. She had an elegantly rounded fantail and a knife sharp bow that sliced cleanly through even the roughest sea. Her two high-power and two low-power turbines could generate more than 27,000 horsepower for turning the two triple-blade propellers, each of which were 9 feet in diameter. The propulsion system could drive the ship to a top speed of 32 knots, or nearly 37 miles per hour.

Captain Hunter was rightfully proud of his accomplishments and stature as a leader and expert navigator, and he prided himself on his mathematical ability and his skill at navigating over long distances in the worst of conditions. To the astonishment of all aboard, for example, he once navigated the mighty *USS Idaho* into the harbor at Anchorage in a thick fog. Leading the 14 destroyers down the coast the next day and night would not be particularly difficult.

Hunter, however, was admittedly distrustful of radio beacons and the newfangled navigation procedures they brought about. He had developed the habit of using the information from these new devices only if it matched with the results of his own calculations based on tried and true methods of navigation and seamanship. Like most sea captains, his strength lay in his strong self-reliance and, at the same time, his adherence to the rules of the chain of command. The *Delphy* was *his* ship, and he was proud having her serve as the flagship for the run down the coast.

With the briefing concluded, the 40 officers returned to their respective ships and prepared for the departure the next morning. Hunter and Watson returned to the *Delphy*, moored just to the port side of the *Melville* in San Francisco Bay, and made their final preparations for the training run the following day. The next time they all talked with each other would be on shore.

The destroyers of Squadron 11 steamed through the Golden Gate the next morning with Commodore Watson in command and Captain Hunter in charge of the *Delphy*. During the first hundred miles of the trip south, the radio beacons from the San Francisco Naval Radio Headquarters matched reasonably well with Hunter's own calculations of the *Delphy's* position. As the hours passed, the squadron moved beyond the range of the beacon from San Francisco, now well behind them to the north. Darkness fell and the sea kicked up a bit. The wind and the current were nearly to stern, and the steersman had a difficult time keeping the ship from yawing and moving off course. The ship occasionally sliced through a large swell, and the propellers could be heard racing as they left and then reentered the water. Captain Hunter spent the hours monitoring the activities on the bridge or talking with Commodore Watson in the adjacent chartroom. When they passed Point Sur, Hunter calculated, using his estimate of their speed and direction, that they were four to five miles off the coast; but it was not possible to see the point and verify their position or their actual ground speed due to limited visibility. They would carry on and pick up another radio beacon from the navigation station at Point Arguello near Point Conception in a few hours. Hunter instructed the helmsman to steer a southerly course of 150 degrees, relying on their magnetic compass because their gyro compass was not operating.

At 1430 hours Hunter instructed the *Delphy's* radio operators located downstairs in the radioroom to request a bearing from the Point Arguello Station to their south. After a minute or so a message came back over the wireless from Radioman 3rd Class G.C. Falls, on duty at the isolated station. Radioman Falls was using a relatively new bidirectional compass that indicated the vector of a radio beam along any two opposing points on the compass.

"Delphy. Your bearing is 167 degrees."

The *Delphy's* radioman relayed the message from Point Arguello into the speaking tube connected to the bridge. Upon hearing the report, Captain Hunter, startling the officers and crew standing about, angrily told the *Delphy's* radioman to tell the operator at Point Arguello to give the opposite or reciprocal bearing to 167 degrees. Hunter knew, as did the others present, that Radioman Falls had mistakenly sent them the bearing from the *Delphy* to Point Arguello, rather than the required bearing from Point Arguello to the *Delphy*. There was a *big* difference between the two possibilities: the first and improper bearing meant that the ship was to the south of the radio station, which it certainly was not; the second and proper bearing meant that the ship was located to the station's north, as it most certainly was. It was a simple and obvious enough mistake, but to Hunter the erroneous report confirmed his strongly held suspicions of navigation by radio bearing. There were just too many chances for error, and he did not like depending on the report from a Third Class Radioman stationed on some remote point of land nearly a hundred miles away.

As Captain Hunter had instructed him, the *Delphy's* radioman returned to the set and called up Point Arguello once again. "Point Arguello, we are to your north. Give us the reciprocal bearing."

Radioman Falls responded by sending the corrected reciprocal bearing of 347 degrees (167 plus 180) which placed the *Delphy* up the coast to his north. With the matter settled, Hunter instructed his officers to continue piloting the *Delphy* on her southerly course with the 13 destroyers following in her wake.

A few hours later Captain Hunter, Commodore Watson, and Lieutenant (jg) Larry Blodgett, *Delphy's* navigator, stood in the chartroom reviewing the situation. Although there was about one mile of visibility on the water, an overhead haze blocked the view of the stars. They still had not seen a light from shore and could not verify their position. Neither could they be certain of their ground speed. They were steaming ahead in the dark in more ways than one, but Hunter, leaning over the chart on the table, reviewed their heading, their time underway, and the most recent count on the revolutions of the propellers. All of this information gave him an indication of their speed and their position.

"We are making very good headway. I judge us to be well off the coast and quite close to Point Conception. We should start thinking about our hard left turn into the Santa Barbara

Channel."

Commodore Watson, who was in a particularly relaxed mood, agreed. "Yes, the sea and the wind have been behind us all day, undoubtedly pushing us along. We have probably made very good progress."

There was a knock at the chartroom door and Hunter responded with a firm "Come in." It was another report from the radioman.

"Sir. NPK reports that we bore 320 degrees true at 1850, sir." This was another radio reading from Radioman Falls at Point Arguello which showed them to be in a position along a 320 degree line extending northwest from the Point Arguello radio station. Hunter thanked the messenger and sent him on his way. The Captain felt satisfied that his earlier handling of the confusion with the radioman at Point Arguello had paid off; the radioman at the station was now providing the ship's bearing *from* - - rather than *to* - - the station.

Hunter leaned over the chart table again and looked at the line he had drawn on the map showing the progress of the Squadron. "All things considered, we appear to be right on course. These radio bearings can be off by as much as 10 degrees, so I would say that we are right where we need to be."

Nothing was said immediately by either Commodore Watson or Navigator Blodgett, but Hunter sensed that the Lieutenant was not entirely in agreement - - and hesitant to say anything at all in the presence of his two superiors who had taken over his navigation duties during the exercise. Hunter, having settled the problem with the 3rd Class Radioman at Point Arguello earlier in the evening, now had to contend with his Lieutenant. He was visibly annoyed when Blodgett began to speak.

"Ah, sir. Perhaps we should stop and take a sounding. We would then know with much greater certainty just how far we are off the coast." Hunter, as well as Navigator Blodgett, knew that the ocean bottom along the California coastline fell off rapidly from the shore. Dropping a weighted line and measuring the depth of the ocean floor would tell them, roughly, whether they were near or far from land. But taking a sounding off the bottom required stopping, and there was no question that Hunter would not take kindly to such a suggestion. The Captain and the Commodore were determined not to interrupt this engineering run and equally determined to be among the first in port in San Diego.

Hunter responded tersely. "Well, I can't see it's necessary. Besides, we would have to break radio silence with the other ships and stop this whole parade. What do you think, Commodore?" Hunter was hedging his bets. He knew Commodore Watson would back him up every step of the way.

Commodore Watson responded as Hunter knew he would. "Yes, ...um, we don't want to stop this show, uh, especially while everything is going so well. I concur. Let's hold the course and proceed as planned."

Navigator Blodgett said nothing further for the remainder of the meeting.

Meanwhile, quiet discussions between captains and navigators took place on a number of the ships in the line, most notably the *Kennedy* and the *Stoddert*. They too intercepted the radio direction messages from Point Arguello to the *Delphy*. All of the bearings were between 320 and 333 degrees, just about the reciprocal of their 150 degree heading, indicating that they were headed directly for the point, although they did not know their distance from it. In direct violation of orders, some navigators aboard the other ships requested and received radio bearings from Point Arguello, all of which confirmed their belief that they were on a course toward the Point Arguello Station. The angle of incidence between their course and the bearing provided by Point Arguello was so small that they could not accurately determine the distance traveled over a period of time through a method of triangulation and simple geometry; so again, they were uncertain how far away they were. At 2000 hours the *Delphy* established radio contact with a very distant Admiral Kittelle on the *Melville* and provided the current position of Squadron 11, a position based on the dead reckoning navigation of Captain Hunter. The reported position of the *Delphy* was considerably further south and further offshore than the position calculated by the navigators of the following ships. The other ships of Squadron 11 intercepted this message, but none challenged it because they had been ordered to maintain radio silence.

At 2030 hours the navigation chart aboard the *Delphy* showed the Squadron to be well south of Point Arguello. The imposing rocks of San Miguel Island lay only a few miles ahead, and to their left were the open waters of the Santa Barbara Channel. Captain Hunter stood over the chart once again with Commodore Watson and Navigator Blodgett. The radioman entered the chartroom to relay a new bearing just received from the Point Arguello Station.

"330 true, sir."

Captain Hunter shook his head in absolute contempt upon hearing what was certainly another impossible bearing. The *Delphy* was now south, not north, of the Point Arguello station. He turned to the radioman. "Tell the station that we are well south of Point Arguello. They are to give us the reciprocal bearing."

"Yes sir," the radioman responded, and turned to relay the message to the radioroom.

The door closed, and Hunter blurted out to Watson and Blodgett: "God, I wish they would get these things straight." He turned back to the chart table and looked at his trace of their progress. For the second time that evening, Hunter concluded, the radio station had given him the wrong bearing. They were now well south of the station according to his calculations, yet the radio bearing from Point Arguello showed them to be to the station's north. The operator had undoubtedly gotten the procedures turned around again and given them a bearing that was 180 degrees off.

Some minutes later Radioman Falls at Point Arguello carried out the explicit orders from Captain Hunter aboard the *Delphy* and radioed the reciprocal bearing to 330 degrees, which was 150 degrees (330 minus 180). The new bearing was reported back to the chartroom where Captain Hunter, Commodore Watson, and Navigator Blodgett studied the map.

Having received the "corrected" bearing from Point Arguello, Hunter concluded that they were finally in a position to turn left and steam through the Santa Barbara Channel. He also knew that they had to execute the maneuver soon or risk running into San Miguel Island, which lay directly ahead to the south on their plotted course on the chart. Commodore Watson concurred with Hunter's assessment and ordered Captain Hunter, according to formal protocol, to "change course to 95 degrees, true" at 2100, just a few minutes away.

Some of the ships in the line were unable to maintain the desired following distance, but most kept up reasonably well. Each ship traveled at 20 knots, about the speed of a sprinter running a 100-yard dash. At this rate each ship traversed the distance between it and the ship in front of it in about 13 seconds.

Captain Hunter and Lieutenant Blodgett entered the bridge from the chartroom, and the Commodore returned to a private room nearby to attend to other business. Ensign John A. Morrow, standing his very first unsupervised watch as the

"Officer on Deck" of a destroyer, stood squarely in the center of the *Delphy's* bridge, forward of the steering wheel and engine-order telegraph. Captain Hunter issued the orders of the change-of-course to Ensign Morrow, who, in turn, gave the order to the helmsman when the second hand on the wall clock ticked straight up to 2100 hours. After two blasts of her whistle to signal a left turn, the destroyer swung to port, cutting a precise white curve in the surface of the dark rough sea. They approached the heading of 95 degrees, or 5 degrees off due east. Captain Hunter walked to the port wing of the bridge to obtain a good view of the line of destroyers following his wake. Through the mist he could just make out the lights of the *USS Percival* which, he knew, was the eighth ship in line. He stepped back inside the bridge and silently monitored the actions of his crew.

About five minutes later, the ship suddenly plunged into a heavy fog. Visibility fell immediately to a few yards, yet the *Delphy* continued relentlessly ahead. Hunter watched as the fog, illuminated slightly by the dim lights inside the bridge, floated in through the open forward windows. All was silent except for the rumble of the ship's turbines and the water spraying across the bow as they cut through the swells. Ensign Morrow stepped forward and stood looking out the open center window of the bridge. Captain Hunter mumbled something about "peasoup."

A strange grating sound worked its way up the ship into the bridge for a second or two. Hunter felt a few quick and fierce bumps beneath his feet. With incredible suddenness the ship ground to a violent and complete stop as it smashed head-on into a solid obstruction. Hunter heard the deafening sound of tons of crashing steel just as he and everyone else on the bridge flew forward into the bulkhead, landing in a heap of bodies and equipment. A few seconds before, they had been steaming ahead at 20 knots, and now they were at a dead stop. Reeling from pain, Captain Hunter extracted himself from the pile of men and equipment and stumbled to his feet and then to the open forward window. The only sound was from the crashing surf. Through the fog he could see the outline of a large and towering rock not more than 50 feet ahead. There was only one answer: they had gone too far south before turning and struck the rugged coast of San Miguel Island.

Hunter, remembering that 13 destroyers followed directly behind, reached for the speaking tube to the radioroom and issued an urgent and panicked order for the Squadron to "Keep clear to westward" and "Nine turn." This was the Navy code instructing the destroyers to make a simultaneous 90-degree turn to the left. Hunter reasoned, based on his mental picture of where they were, that this would direct the other ships in line out to open ocean and into the deeper waters of the Santa Barbara Channel. It did not matter, for the radio antenna had been destroyed by the force of the crash, and his frantic command was barely heard.

The sister ships behind in the fog caught up at a rate of 11 yards each second. Captain Toaz on the bridge of the *USS Lee* looked up in horror to see the lights of the *Delphy* directly ahead. Realizing she was at full stop, he gave the order for full speed astern and left rudder, but it was too late. She slid to the left, and her stern just missed the *Delphy,* whose lights had now gone out.

Hunter, still on the bridge, looked out the port windows and to his horror saw the ghost-like outline of the *USS Lee* only inches away from the rail. The screech and rumble of metal meeting rock was heard on the deck of the *Delphy* as the *Lee* hit bottom and came to a stop.

Just out of Hunter's sight, to the rear of the *Delphy,* the *USS Young* drove hard right behind the *Lee.* She rose up out of the water without warning, running up and over a submerged reef. Volcanic rock ripped her sides wide open, and she rolled over like a beached whale.

The *USS Woodbury*, the *USS Nicholas*, the *USS Chauncey*, and the *USS Fuller* followed the *USS Delphy*, *USS Lee*, and *USS Young* onto the rocks nearby. The *USS Farragut* backed into the *USS Fuller*, ran aground, and then managed to back off to deeper water, as did the *USS Somers* after hitting a reef and taking in water. The *USS Percival*, *USS Kennedy*, and *USS Hamilton* managed to reverse course and avoid the other ships and rocks when they heard and saw the commotion immediately ahead. The *USS Thompson*, the last of the 14 ships in line, stopped when the ship in front of her turned left. Depth soundings were taken, and she turned westward to deeper water, not fully aware of the unfolding tragedy one and a half miles to the east.

Heavy surf broke over the seven stranded destroyers, eventually breaking the *Delphy* in two. Hundreds of thousands of gallons of fuel oil from the ships spilled into the crashing waves, carbide bombs exploded when they came in contact with the water, and numerous fires began on the ships and on the surface of the sea. Twenty-three men died in the ensuing attempts to abandon the ships and make it to shore in the darkness and heavy surf.

The 450 survivors from the seven destroyers made their way slowly to the narrow and rocky beach below the cliffs during the hour that followed. Two hundred of the survivors were badly injured, many burned, but they were destined to wait out the night until rescue, perhaps in the morning. Some scaled the cliffs above the beach to seek out shelter and possible help, while most gathered on the narrow strip of shore at the base of the cliffs. Among those gathered between the crashing surf and the cliff were Captain Hunter and Commodore Watson, standing among a group of wet and shivering ship captains and officers.

Their prospects for immediate rescue did not look good, especially considering that their probable location was the northwestern edge of desolate and remote San Miguel Island, 40 miles south of Point Conception. Commodore Watson, trying his best to address the problem at hand, assessed the situation with Captain Hunter and the other officers who made it to the rocky windswept beach. Captain Hunter, visibly shaken but still in control, stood next to the Commodore, listening as he issued general orders for moving everyone up the cliffs and establishing some sort of camp to wait out the cold night.

Hunter thought he heard something, and, a second later, the others in the group stopped talking. Yes, there was something out there. A strange whistle sounded again in the distance. It seemed to be coming from off down the coast, but, oddly, from up above as well. The whistle sounded again and a few of the enlisted men standing about shouted with excitement upon hearing it, for they knew then that they would soon be rescued. To Captain Hunter, however, it meant far more than that. Up on the bluff above the beach a train was approaching - - a train of the Southern Pacific Railroad. Captain Hunter stopped shivering and only then realized that he was standing on the beach just around the corner from the Point Arguello Radio Station along the central coast of California.

REFERENCES AND NOTES

Casey, S. (1993). That newfangled technology. *Ergonomics in Design*, January, 33-37.

Driever, D. (1992). Destroyers down. *Naval History*, Spring, 20 - 25.

Lockwood, C. C. and Adamson, H. C. (1960). *Tragedy at Honda*. New York: Chilton Company.

Overshiner, E. E., (1980). *Course 095 to eternity: the saga of destroyer squadron eleven*. Santa Rosa, CA: Elwyn E. Overshiner, Publisher.

Steiger, G. (1983). Survivors to mark 60th anniversary of naval disaster. *Santa Barbara News-Press*, August 21, D-1.

Sweetman, J. (1984). *American Naval History*. Annapolis, Maryland: Naval Institute Press.

IN SEARCH OF THE LOST CORD

Karen Nessen checked the EKG electrodes taped to the chest of the four-year old girl and ran the lead across the starched white bed sheet to the upper corner of the mattress. All she had to do now was plug the lead into the cord from the heart monitor machine sitting nearby at bedside. Karen and the other staff would then be able to monitor the little girl's condition while they were down the corridor at the nurse's station. An alarm would sound if the machine detected a significant change in the rhythm or pace of her heart.

Karen, like the other nurses and technicians there at *Children's Hospital* in Seattle, treated dozens of young patients every day. Some of the cases were simple. Some, like this little girl's, were complex. But they all brought forth the extremes of emotion one has when treating the young. One minute you experienced the joy of helping a sick or injured child regain her health, and the next you might suffer the pain that accompanies failure and loss of hope.

This particular little girl had spent much of the four short years of her life combating a number of significant birth defects. She had fought many difficult battles in the past and would have to fight many more in the future. Everyone at the hospital was determined to provide her every opportunity to overcome her handicap.

Karen pulled the sheet up and folded it back neatly over the girl's small chest. She looked much more comfortable now that everything had been taken care of and the bed was straight. Not one to miss the details, Karen then reached down to the guard rail on the side of the hospital bed and lifted it upward. The metal bars made a satisfying click as they locked into place. There would be no danger of the patient falling out of the bed as she slept.

All that remained was to plug the lead from the electrodes taped to the little girl's chest into the cord from the heart monitor machine. Karen picked up the lead resting near the corner of the mattress and looked down for the connecting cord. She paused for a moment to sort things out. Another machine was next to the heart monitor at bedside. It was a portable intravenous (IV) pump with all of its assorted cables and lines.

Ah! *There* was the cable, hanging down near the side of the heart monitor. She bent over and grasped the end of the cord with her right hand and brought the two pieces together to connect them. The cord in her left hand terminated with a circular six-pin connector frequently used on EKG leads. It had to be held in the proper orientation when plugged into the receiving end of the cable from the heart monitor. It wouldn't fit into the receiving end unless it was held in just the right way. Karen lined up the two ends and started to push them together.

There was not even a remote possibility in Karen's mind that the cord in her right hand was from anything but the heart monitor machine. After all, she handled dozens of small machines every day, and they all seemed to have different types of connectors. These unique connectors certainly made it easier to find the pieces that were supposed to fit together. More important, it made it nearly impossible to connect two things that were never meant to be connected. Or such was the intent.

Little did she know that the cord in her right hand was from the nearby IV pump and not from the heart monitor. Its size and shape were similar to the cord from the heart monitor, and the connector had slots that matched reasonably well with the six-pin termination on the EKG electrode lead taped to the child's chest. The similarity might not have been so critical had the IV pump not been one of the portable models that could run off a battery or wall current. Accordingly, it had a detachable power cord for portable battery-powered operation. It was this cord that she had mistakenly grasped. Karen had no means of knowing that the cord in her right hand was a live electrical circuit and that it carried the full operating current of the IV pump!

In that dreadful moment Karen Nessen plugged the two connectors together and a lethal current of electricity streamed through the cord of the IV pump, through the errant connection, down the lead and one of the electrodes attached to the girl's chest, through her heart, and back to the IV pump via the other electrode and lead. The reaction was immediate, and Karen separated the two cords within moments of realizing what had happened. Karen and other members of the staff administered cardiopulmonary resuscitation to the girl during the terrible minutes that followed, but they could not revive her. A hospital spokesman stated to the press shortly thereafter: "At this time it looks like this is just a human error which is fairly tragic."

REFERENCES AND NOTES

Hospital electrocution of child is investigated (1986). *New York Times,* December 5, 16.

Neumann, P. (1987). Illustrative risks to the public in the use of computer systems and related technology: Plug-compatible modules and an electrocution (with comment from Paul Nelson). *ACM SIGSOFT Software Engineering Notes,* 12(1), 10.

Nurse's error causes girl's electrocution (1986). San *Francisco Chronicle,* December 4, 29.

A pseudonym has been used for the main character in this story. Any similarity to her real name is purely coincidental. The thoughts and observations attributed to the main character are fictitious, but are believed to be representative of the setting and circumstances. The major technical elements of this story are based on published accounts and are believed to be accurate.

AN ACT OF GOD

William "Bill" Jurith, System Operator, reached for the ringing telephone on the console in Consolidated Edison's Power Control Center at the corner of West End Avenue and 65th Street in Manhattan. The caller was probably the System Operator at Orange and Rockland Utilities to the north. The displays on the control board in front of Bill were giving off the telltale signs of a problem with one of the main electric feeder lines coming into the city; the lights had begun flashing just moments before. Bill was acutely aware of his direct responsibility for delivering electricity to the occupants of New York City on that muggy evening in July. His job was to juggle the supply of electricity with power consumption, and any disruption in the flow from the north was of concern. After all, there were only a handful of high-capacity ties to the island. The loss of even one was important.

Bill, 56 years of age, commuted into the city each work day from his pleasant, green two-story house in Brooklyn where he and his wife raised their family. He took the job of System Operator about three years before when he accepted the promotion from his previous position as Power Dispatcher. Like many other senior operators in electric utility companies, Bill was chosen for his new job primarily because he had amassed considerable experience with the complex electric network during his years of service to the company. True, the new job was considerably more challenging and he had undergone treatment for a mild case of hypertension since starting, but the condition had not interfered with his performance during the normal work day.

But due to circumstances beyond his control, this was not going to be a normal shift in any sense of the word. Bill would soon find his skill, knowledge, and demeanor tested as never before. His actions at his console during the next hour would affect the lives of over eight million people in the City of New York.

Bill put the receiver to his ear. The operator from the utility got right to the point: "Severage."

Bill responded in the laconic vernacular used by utility technicians, technicians who work with each other day in and day out. He asked him if the feeder line, line Y88, had "opened" (stopped carrying power) as his control panel suggested: "Yeah, Y88 opened up?"[1]

[1]This dialogue, like all of the dialogue in this story, is taken from an actual tape recording of telephone conversations made the night of this event. Only selected conversations have been used in this story due to the length of the actual transcript.

"Well, this is what I am assuming," the operator replied. "I haven't heard a thing. I was just trying to get a hold of him before I called you to confirm the thing was off." The operator was talking about yet another operator in a distant facility who would know something about the status of the important power line.

Bill looked at his panel for the Y88 line and saw that the big circuit breakers up north had stopped transmitting power to New York City. "I see the breakers are opened up there." An "open" breaker meant that the circuit had been disconnected and electricity was not flowing.

The utility operator was obviously looking at his own control board as well: "Yeah, OK." They both hung up. It was a typically brief conversation.

One minute later at 8:38 p.m., less that two minutes after the "opening" of the Y88 line, Bill was into his third telephone conversation, this time with the Con Edison Westchester District Operator located in another control room next door. Bill learned that lightning was seen up north and that it might have hit the Y88 line.[2]

---

[2]What he did not know was that a severe lightning stroke hit two 345,000 volt (345-kV) lines called the W97 and W98. These two transmission lines ran together on the same tall towers, and the lightning bolt was sufficiently powerful to put both of them out of action. Four circuit breakers were triggered, as designed, but it resulted in the automatic shutdown of the Indian Point No. 3 nuclear plant located on the Hudson River some miles north. The nuclear plant, having no place to send its generated power, tripped off-line, thus reducing the electric flow toward New York City by 833 megawatts

Fortunately, redundancies and back-up systems made his job easier. For example, there were six major lines feeding New York City. Bill's primary responsibility was to maintain the supply of electricity to the city. If one of the six major feeder ties went out (by, say, a lightning stroke such as this), the remaining ties could take up the necessary load. Many of these recovery functions were provided by other technicians or automatic systems, but redistribution would require some quick action on Bill's part. The second action that had to be taken during a power loss was to increase in-city power generation, something for which Bill had direct responsibility.

But much of Bill's own time and energy during the next eight minutes was spent talking to other operators on the network about the distant power lines. His console was not up-to-date, and it did not present overall, supervisory information about the status of the sprawling system. He needed a global understanding of its condition in order to perform his job, and in this case he had to figure much of it out for himself. Even so, his conversations with other operators were very general and meandering, and he failed to learn anything of any real use for the task at hand.

(MW). The loss of the W97 and W98 lines also resulted in the isolation and loss of the Y88 Ladentown line, a 347-kV tie to the city. The combined loss to Con Edison was a whopping, but recoverable, 1,310 megawatts. The power flow over the five remaining ties to Manhattan increased to make up the loss of the power from the tripped lines and open tie. (The network power lines are referred to as "lines" and the six connections to the island of Manhattan are referred to as "ties" throughout this story.)

Eight minutes and eight telephone calls after the line failure up north, Bill's phone rang again. It was Bill Kennedy, the Senior Pool Dispatcher at the New York Power Pool in Guilderland, New York. Kennedy was calling to check up on Bill and to make sure he was increasing in-city generation.

Bill answered the phone and identified himself. "Yeah, Bill."

Kennedy was polite but direct. "You getting your turbines on - - eh, Willie?"

"Yes."

"Yeah, OK, fine..." Kennedy eased out of the conversation and left Bill alone to carry out his important work.

Within a minute, Bill sent out a request for reserve in-city generation to help replace the loss caused by the lightning stroke. This required the stations within the city to increase their output immediately and to provide some level of reserve in the unlikely event of another transmission failure. It was something Bill should have done eight minutes before, but he was busy on the telephone. He also was not fully aware of the magnitude of the power line outages up north.

At 8:50 p.m. Bill believed he had taken the appropriate steps to address the problem, albeit somewhat slowly. The load on the major ties leading into New York City had been increased. Also, the in-city generating capacity was on the rise. The situation was disconcerting, but certainly still manageable as the top of the hour approached.[3]

[3]Bill was not aware of the precarious state of the entire system. First, the power output from four of the five in-city steam plants was coming on much more slowly than planned. Like Bill's request to increase power, the response of the in-city generating stations was very slow, and only one, the Arthur Kill Steam Station, was actually generating as much power as required. Second, the energy combustion turbines stationed throughout Manhattan were not being started as quickly as needed. One group of turbines was even out of service due to

Fate struck again at 8:56. A powerful bolt of lightning hit a tower supporting a major connecting line upstate, putting two 345-kV lines out of service. This loss, in turn, resulted in the isolation of another northwestern tie to Con Edison, the 1,044 MW Y94 Ramapo tie. When it failed, a resulting power surge caused the W81 line leading down from the north to trip out as well. All of these line losses meant that two of the six ties feeding power to New York City were now out of commission. The four remaining ties were overloading as they took up the slack to meet the power demands of the city. There was still hope of recovery, however, but it required that Bill increase in-city generation, balance the load of imported power over the four remaining ties, and institute a reduction in power supplied to the city.

But due to the sketchy information presented on his control panel and, perhaps, to his predisposition and hesitancy, Bill failed to understand that two of his major ties to the city were out of action. Specific transmission-circuit in-service status was not displayed explicitly on his panel, and it was up to him to decipher the power flow data and determine which, if any, lines were down. Unfortunately, his conclusion that it was sufficient to spend his time continuing to talk with other operators about the status of various lines was entirely incorrect. He had to act

---

maintenance work, something that should have been reported to Bill long before. In all, the in-city generating stations were providing 341 MW less than they, in combination, told the control center they could produce. Third, and perhaps most importantly, Bill did not understand the status of the entire network and its current vulnerability. His control panel consisted mostly of hundreds of independent displays, each associated with an important network component.

in order to avoid a possible series of line losses, line losses that could rapidly escalate into a system-wide failure.[4]

His 13th call in 18 minutes was from Kennedy, the Senior Pool Dispatcher at the New York Power Pool. It was the third time Bill talked to Kennedy that evening.

Bill responded "Yeah" when he picked up the receiver.

Kennedy warned him that he was in danger of losing even another tie if he failed to increase the city's own generation, lighten the load, and get power from somewhere else. "Bill, you'd better get that Linden back or you'll lose that baby too." Linden was one of the important ties into the city. Kennedy was telling him that he had better lower the load on the Linden/Goethals line or it would blow. The Linden tie brought power in from the west, down below Manhattan, across Staten Island, over to Long Island, and then back westward to New York City.

"Yeah, I just lost 81." Bill was referring to one of the lines up north that he lost moments before.

"Yeah."

"Yeah," responded Bill.

"You lost the Jamaica tie, too," continued Kennedy.

"Which Jamaica tie?"

"Ah, Jamaica Valley Stream," said Kennedy. Another tie into Manhattan had ceased transmitting, but only temporarily. This was the first Bill had heard of it.

---

[4]The control board for the Westchester District Operator, located in the room next to his, had a more modern graphic display of the network. Lights flashed and a siren sounded, telling the operator that the ties to New York City had failed. But this was of no use to Bill, as he could not see the lights or hear the alarm from his control room.

"Oh, that could be. Alright, but, ah, help me out with the 81/80 feeder, huh?"

The situation had been mildly stressful since it began 25 minutes before, and he knew he had been a little slow getting in-city generation up. But this was the first time Bill let anyone else know that he might be in over his head. The demands on him had increased considerably during the past minute, and his request for assistance from someone who was not necessarily in a position to help him was the first outward sign of the difficulties he faced.

"Yeah," said Kennedy.

"Right," responded Bill, prior to hanging up the receiver.

Bill's phone rang again less than a minute later. It was Kennedy again.

"Bill, you better shed some load until you get down below this thing because I can't pick anything up except from the north, see?"

"Yeah."

Kennedy continued. "So you'd better get - - do something to get rid of that until you get yourself straightened out."

"I'm trying, I'm trying," he snapped back.

Bill fielded two additional telephone calls during the next 60 seconds - - one from a system operator at the Long Island Light Company and another from the Con Edison District Operator at Brooklyn and Queens. Meanwhile, the lines serving the ties to New York City began to shut down from overloads, like a long line of teetering dominos. The load on the remaining lines increased with each successive failure. All the while, Bill presumed that his major tie to the north, the one lost during the second lightning strike, was still available. He continued spending much of his time talking to other operators on the telephone.[5]

The phone rang at 8:58. It was Kennedy.

"Bill, I hate to bother you but you better shed about 400 MW of load or you're gonna lose everything down there." Kennedy's tone of voice had changed from their last exchange. He was slightly sarcastic and much more direct.

Bill said the first thing that came to mind. "Yeah, I'm trying to."

"You're trying!" Kennedy popped back. "All you got to do is hit the button there and shed it and then you worry about it afterwards, but you got to do something or they're going to open up that Linden tie on you." The big Linden tie was one of Bill's hopes for getting out of this mess. It too was becoming severely overloaded, and it was usually preferable to have a controlled shutdown of a line rather than let it blow by itself.

"Yeah, right, right," said Bill.

Two minutes later they were on the phone again. This time Bill called Kennedy, signaling a change in circumstances and Bill's renewed awareness of the growing severity of the

[5]During none of the 16 telephone conversations since the first lightning stroke had anyone explicitly told Bill that his main northern tie, the second one to go, was out of service. If he had known this, he could have flipped a switch on his panel to reset a breaker. This would have restored power over this major tie to the city.

situation. It was a call for help.[6]

There was a strange urgency to Bill's plea. "Does it look any better?"

Kennedy's response was immediate and direct. "No, you still got to get rid of about 400, Bill, because you've got - - you're 400 over the short-time emergency on that 80 line."

"Yeah, that's what I'm saying. Can you help me out with that?" Bill was asking for some help with that northern line that he mistakenly believed was in service.

Kennedy's response was harsh. "I can't do nothing because it's got to come from the lower part of the state and there's nothing there to help you with. You got to do it...."

Kennedy's words were downright unfathomable. Perhaps he didn't have all of the facts. Bill shot back: "I got no GT (Gas Turbines) to put on 'cause they went home." Bill was referring to the slow start-up of the in-city gas turbine generators due to the slowness of his own actions, the unannounced maintenance work, and the generally slow response time.

Kennedy laid it all on the line for Bill. "OK, then you're gonna have to shed load, because that's the only way that thing's

---

[6]Bill was still under the impression that the big feeder line up north, the Ramapo tie, was available to route power down to him. It alone had more than enough capacity to carry the load from the New York Power Pool down to Con Edison's system. All he had to do (in his own mind) was get the Pool Dispatcher to reroute the load down to the available tie. But without explicit information, together with the continued diversions on the telephone, it never occurred to Bill that the line was out of commission. Furthermore, Bill was never explicit in his request. At the same time, it had not occurred to the Senior Pool Dispatcher that Bill didn't know that this line was unavailable. It was such a fundamental part of the whole problem. Bill and the dispatchers just talked around the problem, and Kennedy never really answered Bill's question. Instead, he continued telling him what to do.

gonna save you."

"Yeah," acquiesced Bill.

"So you get those.... things on," he said, referring to the Gas Turbines.

Yeah, right," responded Bill.

"I told Long Island to pick up everything that he has. That's the only place I can get into you."

"Anything you can do to help?" pleaded Bill again. It was as if he had not heard a thing Kennedy had just said.

"There's no way I can help you, see. OK, Will?" shouted Kennedy as they cut off.

It was now 9:04 and the situation was deteriorating rapidly. Yet Bill, overwhelmed with his predicament, took no action to shed load within the city, and continued to talk about the tie losses with various operators in the network. In the span of just under 30 minutes, the power grid supplying New York City had gone from a functioning state to the brink of total collapse. One of Bill's last remaining ties to the outside world, known as Feeder 80, was drastically overloaded, and he called the Westchester Emergency Supervisor to talk about cutting out the 80 line. The facility was not fully staffed at that late hour, so Bill and the Emergency Supervisor decided that Bill could cut out Feeder 80 himself from his control room if that is what he wanted to do.

But before he made the final decision, Bill called Kennedy once again. He got right to the point. It was now 9:05.

"I'm going to cut Feeder 80. I have no way of deloading it right now. I'm gonna cut it out."

Kennedy was equally direct. "Can't you shed load and

relieve it? If you cut Feeder 80, then you'll really be in trouble!"

Bill reconsidered. "Ah, I'll see what I can do."

Bill's options were diminishing each minute he failed to act. He had four principal problems. First, it was still not entirely clear to him that two of the six lines to the city - - both of the lines from the northwest - - were out of commission. Second, he wasn't getting any help from the north or northwest (for reasons he did not fully understand), and, based on what Kennedy had to say, he couldn't expect any. Third, his Feeder 80 tie was seriously overloaded. If he cut that link, the loads on the other ties would increase even more. Fourth, his in-city generators were still painfully slow making up for the loss of imported power.

At 9:08 he could think of just one thing to do: call his boss at home. He was Charlie Durkin, Con Ed's Chief System Operator.

Bill introduced himself as Charlie picked up the receiver at home. "Hello, Bill."

"Hi ya, Bill," responded Charlie. Charlie was sitting at home under the light of a kerosene lamp. He had lost power to his neighborhood some minutes before, so he had some forewarning of Bill's problems, but certainly no awareness of their magnitude. "You got some problems, huh?"

Bill gave Charlie a hurried but rambling run down of his "problems." It was a frantic summation. "Yeah, Charlie. Just one moment, huh. I got - - I lost Y88 and W98, and it looks like W97 is alive back on one breaker from Millwood. Indian Point - - then I got overload, and I got 81 taken down, but it must have been struck by lightning - - because 81 went out. I'm overloaded on 80 by 1,430 megawatts. I'm trying to get everybody back up but I have no GT's. I had......."

"One thousand four hundred thirty MWs on 80?" exclaimed Charlie.

Bill could depend on old Charlie. He was a young, bright engineer and had an expansive understanding of the system. Not surprising, within a minute Charlie had deduced that the big Ramapo Y94 tie from the northwest was out of commission. This was the first time anyone had communicated this directly to Bill. He had not determined it himself before due to the sketchy information on his control board, his knowledge of the system, and his own powers of deduction under the severe stress.

The hour was getting late, but Bill spent all of the next ten minutes bringing Charlie further up to date on the worsening situation. It was difficult to think clearly under all of the pressures. Bill's comments were confused and segmented. At first he told Charlie that he wanted to sever the 80 line because it was overloaded, but, as Kennedy had said earlier, this would only compound the problem by increasing the demand on the remaining ties and in-city generation stations. Bill listened as Charlie told him that cutting the 80 tie would only make matters worse. Charlie made it clear that Bill had to reduce the electric load within the city. At 9:09 Charlie started to tell Bill to go into voltage reduction, but then Bill interrupted him.

"You either go into voltage..."

Bill broke in in a panicky voice. "And the Linden tie is way up there. I had to back off, because he was up to 900 MW; the guys down there are about 500."

Things were spiraling out of control.

They continued to discuss line loading until 9:13, at which time Bill finally admitted for the first time that they had no alternative but to reduce power output. "We've got to get more off; we're going to have to go into voltage reduction."

Charlie agreed wholeheartedly, and Bill began to operate the panel in order to reduce voltage five percent. It took him five long minutes.[7]

At 9:17 Charlie told Bill that the five percent voltage reduction was obviously not enough. He would have to go to eight percent reduction. Bill began to initiate these changes on his panel, but it was a slow and tedious process.

At 9:19 there was another line failure up north. This time a 345-kV line, extremely overloaded for the past 23 minutes and stretching low toward the ground, tripped out when it contacted a tree. It had been supplying 1,202 MW to the ties leading to New York City. The four remaining ties to Manhattan were importing 1,900 MW of the load that was not being supplied by in-city generating plants. One of the lines, the Linden/Goethals tie, was now carrying a load of 980 MW, well above its rated load of 505 MW.

A few tenths of a second later, one of the remaining ties from the northeast tripped out from the overload (the Pleasant Valley 345-kV 80/81), leaving three main ties to carry the power into the city. As before, each remaining tie was loaded with the current that was previously carried by the lost tie. The Linden/Goethals tie was now transmitting 1,090 MW, more than double its normal rating; the Jamaica/Valley Stream tie from Long Island was up to 310 MW, above its normal rating but below its emergency rating; and the last tie from New England was carrying 340 MW, well above its emergency rating of 250

[7]While Bill was on the telephone with his boss, Bill Kennedy, the NYPP Senior Pool Dispatcher, was trying frantically to contact him to coordinate corrective actions. The situation was still salvageable, but it required close coordination and load shedding within the city. But he could not get through because Bill was still on the telephone.

MW.

At 9:22 the operator at the Long Island Lighting Company, unable to contact Bill, manually cut out the Jamaica/Valley Stream tie to Manhattan after it reached a dangerous load of 520 MW. It was a prudent move, for the Long Island power system was in danger of being pulled down with the Con Edison system. But this left only two of the original six ties to New York City.

The whole thing had avalanched beyond belief. Each failure led to two others, and each of these led to two others in turn. The entire system was crashing down. They had no choice but to start cutting off sections of the city that still had power. Charlie instructed Bill to begin the procedure, but Bill hesitated, as if he did not know what to do. He had rehearsed the operation years before and knew the procedure in the manual, yet under the stress of the moment it all seemed so confusing. Accordingly, Charlie tried to step Bill through the task over the telephone. It wasn't all that difficult.

All Bill had to do was activate the system and press the "Area Trip" pushbutton for a given station. But when he reached for the selector dial to start the process, Bill turned it the wrong direction - - toward the "Frequency Control" position rather than to the "Trip/Reclose" position! The "Frequency Control" position was a relatively unrelated function, yet it was controlled by the same dial.[8]

Bill then proceeded to press the "area trip" buttons for the various stations, but nothing happened.

"It won't trip the son of a gun..." he pleaded to Charlie through the telephone.

[8]This analysis is based on information provided on page 33 of The Con Edison power failure of July 13 and 14, 1977, final staff report (June 1978). U.S. Department of Energy, Federal Energy Regulatory Commission.

Charlie kept encouraging him. "Just keep going. Give it a shot - - if it won't go, go on to the next one."

"What the..."

With the power demands vastly exceeding supply and Bill unable to black out sections of the city to relieve the electric demand, the loads on the two remaining ties to Manhattan continued to mount. At 9:29 the Linden/Goethals tie, now carrying 1,150 MW of power to Con Edison, failed. At that point the total generating deficiency of 1,680 MW was placed on the remaining ties from Pleasant Valley. They tripped out immediately. Manhattan was now isolated from the outside power grid.

Bill was unable to shed load from his panel because the selector switch was in the wrong position ("Frequency Control" instead of "Trip/Reclose"), and the regional in-city generating stations were unable to meet the power demands of Manhattan. At 9:36 p.m. the Con Edison electric system collapsed and Manhattan plunged into total darkness. Power would not be restored for 25 hours. The unequalled failure was, in the words of one Con Edison official who spoke to the press the next day, due to "an act of God."

EPILOGUE

The U.S. Department of Energy's Federal Energy Regulation Commission's investigation into the calamity concluded that "The single most important cause of the July 13, 1977 power failure was the failure of the system operator to take necessary action. The Con Edison system could have survived the lightning-induced line outages, the loss of the Indian Point No. 3

generator, and the equipment malfunctions that extended the impact of these events. It could not have survived these events, however, without either some load shedding or prompt increase in in-city generation. The system operator, the only person authorized and able to take these actions, did not call for increased in-city generation at the proper time and did not initiate manual load shedding until too late, although repeatedly advised or directed to do so and even when that action became the last hope of averting a major interruption." In addition to making extensive recommendations regarding lightning protection and transmission system design, the Commission made numerous recommendations pertaining to the design of the control room, automated decision-aiding systems, communication, training, and selection of operators. Con Edison implemented most of these recommendations, as well as the recommendations made by internal investigators, the City of New York, and the State of New York.

REFERENCES AND NOTES

Beame cites Con Ed 'omission' in blackout report (1977). *The New York Times,* August 10, IV-16.

Boffey, P. M. (1978). Investigators agree N.Y. blackout of 1977 could have been avoided. *Science,* 201, September 15, 994-998.

Carey allots $1 million from state to hire 2,000 for blackout cleanup (1977). *The New York Times,* July 27, II-4.

Con Ed, an hour after the blackout started, tried a wide restoration of power in vain (1977). *The New York Times,* July 28, II-9.

Con Ed controller, citing ill health, not at hearings; investigator calls his pre-blackout action 'panicky' (1977). *The New York Times,* August 31, IV-13.

Con Edison admits its operator erred (1977). *The New York Times,* August 29, 1.

Con Edison now lays blackout failure by man and machine (1977). *The New York Times,* August 25, 1.

Con Ed's pre-blackout calls punctuated by worries (1977). *The New York Times,* August 26, II-2.

Controllers differ on blackout fault (1977). *The New York Times,* September 1, III-8.

Heart of darkness (1977). *Newsweek,* July 25, 16-31.

Perrow, C. (1984). *Normal accidents: living with high-risk technologies.* New York: Basic Books, Inc.

The Con Edison power failure of July 13 and 14, 1977, final staff report (June, 1978). U.S. Department of Energy, Federal Energy Regulatory Commission.

The control room that didn't control (1977). *The New York Times,* September 10, 24.

State study on blackout criticizes Con Ed Control-room conditions (1977). *The New York Times,* September 17, 27.

Wilson, G. L. and Zarakas, P. (1978). Anatomy of a blackout. *IEEE Spectrum,* February, 39-46.

TIGERSHARK!

The Tigershark turned its bullet-like snout toward the open space ahead and slowed to a stop. Its gun-metal gray skin, illuminated by sun-fired rays, pulsed with energy. This predator was the product of relentless evolution, bred to survive in a hostile environment, destined to kill. She was among the most threatening of her kind, capable of sensing prey from great distance, stalking without detection, and slaying with blinding speed and precision. But unlike her living namesake that roamed the depths of the ocean, this Tigershark was a machine, designed to hunt in the expanse of air and sky. Her environment was the demanding and threatening arena of the high-technology jet fighter and, once airborne, she would prove herself to be one of the most capable combatants in the world.

David Barnes, Engineering Test Pilot, Northrop Corporation, sat in the cramped cockpit of the F-20 Tigershark Prototype GE-1001, USA civil registration N3986B, waiting for clearance to begin his roll down runway 08 at the Goose Bay Airport in Labrador, Newfoundland on a remote stretch of the North Atlantic coast. The long gray nose of his fighter pointed down the 11,050 foot runway. Barnes, having completed all of the preflight checks, rehearsed the five-minute flight in the back of his mind, picturing the movement of the Tigershark as he put her through her paces high above the runway. He visualized and felt every pull on the hand control and every press on the rudder pedals, and he imagined the forces of gravity tugging on his body as he powered through the turns and pulled out of the dives. He had performed the five-minute set of intricate maneuvers 60 times before during the preceding weeks, the last rehearsal having ended just a few minutes before. Now, the aircraft had been refueled, and he was ready to do it again.

Barnes was thrilled to be flying this machine, especially through an exceptionally demanding aerobatic scenario such as this. The five-minute routine he was about to rehearse would demonstrate the capabilities of the F-20 to the world, so it was imperative that she be flown right up to her limits. But Barnes, representative of his profession, knew that with this new generation of fighters it was the pilot - - not the aircraft - - that defined the bounds of flight. The Tigershark could take far more punishment than he. It took all of his attention and physical skill to keep the plane under tight control, always with full precision and always while taking care of the hundred other things he had to do in the cockpit. One slip could bring his life to a sudden end. But David Barnes had every intention of living to see his 41st birthday so, as always, he would apply every ounce of his mental and physical abilities in the following minutes.

In just a few days Barnes and his friend and fellow F-20 test pilot, Paul Metz, would be on their way to France for the greatest spectacle in aviation, the Paris Air Show. Traveling with them was their ten-man support team of motion-picture photographers and maintenance personnel, as well as the folks to take care of everything else necessary to get an F-20, a Gulfstream II business jet, and tons of equipment all the way from Southern California to Paris. It had come together amazingly well, and these last few days of rehearsals before making the final cross of the Atlantic provided a great opportunity to put the finishing touches on the show.

Barnes, like the other members of the team, had high expectations for the trip and for the Tigershark. Developed at a cost of over 1.2 billion dollars and without any direct funding or orders from the federal government, Northrop strategists planned on a large and lengthy production run based mostly on foreign sales. No foreign buyer, it was reasoned, could pass up a fighter with 90 percent of the capability of an F-18 at the bargain basement cost of 15 million dollars, just a fraction of the price of comparable fighters from the United States and Europe. Early tests of the prototypes had proven so successful that the company had recently approached the Air Force with an offer to sell them 396 F-20s over a period of years, and Northrop would toss in all support equipment for no additional charge. Barring unforeseen events or circumstances, the F-20 was soon to be a commercial success as well as a technological triumph.

Barnes surveyed the scene out the bulbous window surrounding his head. The air outside was a clear and cool 38°F, and a bit of snow remained on the ground from the last winter storm. The sky was clean and blue and the horizon was generally flat and well defined, providing ideal visual contrast between sky and ground for his dangerous turns and rolls. It was imperative that he keep the horizon line in sight at all times, especially considering his breakneck speed and low altitude.

Just as he completed the mental rehearsal of the flight and survey of the surroundings, the Canadian tower controller radioed that he was cleared to take off. With that signal, Barnes pushed the throttle lever in his left hand forward to the full afterburner position, let up on the brake pedals, and accelerated down the 200-foot-wide runway sitting nearly atop the General Electric YF-404-GE-100 jet engine. The flames roared out the back of the jet and propelled the 16,500 lbs of aircraft, fuel, and David Barnes down the runway at blazing speed. This was the 400th flight of this F-20 Tigershark prototype, and the hour was 1:45 p.m., Atlantic Daylight Time, May 14, 1985. Barnes would complete his planned maneuvers and be back on the ground in only five minutes.

The sensation of accelerating down a good runway in a jet fighter was unmatched by any other, and Barnes had experienced it many times before during 120 combat missions in Southeast Asia and while piloting a variety of military aircraft, including T-38s, F-4s, F-15s, and F-111s. He had gained a great deal of experience during his years in the Air Force, most of it in the service's workhorse fighters. Once airborne, however, none of these older planes matched the sensation of piloting the new F-20. She could enter a turn or pull up from a dive at high speed, subjecting the airframe and the pilot to rapid-onset gs. In the span of a single second, the F-20 was capable of transforming Barnes' suited weight of 200 lbs, including his boots, helmet, and standard-issue g-suit, into a mass weighing well in excess of 1,000 lbs. Furthermore, the F-20 would continue to hold the turn, pull up, or roll for just as long as the pilot could hold her in that position.

As Barnes pulled up from the runway and raised the landing gear, he felt his g-suit tighten around his legs and abdomen. Although it was a common standard-issue Air Force garment, it was the product of decades of research aimed at combating the deleterious effects of the artificial gravity created in high-performance aircraft. This was Barnes' fourth flight of the day, and the garment was beginning to be quite uncomfortable, but it was an essential piece of his equipment and had to be worn. In addition to the numerous tight zippers and laces up and down his legs and abdomen, this anti-g coverall had long, tube-like air bladders running the length of his pants. These interconnected bladders were attached to a hose that ran from his suit into a port next to his seat in the cockpit. When he was not accelerating or "pulling gs," the bladders remained deflated, but as the plane's sensors detected downward forces, as when he entered a banked turn or pulled up from a dive, pressurized air was pumped through the hose and into the bladders. The garment, especially the lower sections around the legs and stomach, tightened dramatically, just like the cloth cuff wrapped around a patient's arm during a simple blood pressure check. The result was that the blood in Barnes' body, instead of pooling down in his torso and legs, stayed distributed through his upper body and head where it was needed most.

The g-suit enabled him to fly maneuvers that he would otherwise not be able to perform. *Without* the suit, the blood in his eyes and head would be forced down into his legs and feet. *With* the g-suit, he extended his range of operation from three or four gs (three or four times the force of gravity on earth) to seven, eight, or more gs. This was crucial during particularly demanding maneuvers, such as those a pilot encountered during an aerial dog fight. The g-suit enabled a pilot to turn faster and harder and obtain the advantage over an otherwise equal opponent. Still, this flight was no piece of cake, and it required an exceptional pilot to pull it off.

From take-off to landing, the five-minute routine was designed to simulate aerial combat and demonstrate the capabilities of the F-20. The flying was fully choreographed, and Barnes proceeded to take the plane into eight consecutive maneuvers, each one designed to exhibit the plane's capabilities and to thrill the audience on the ground in Paris. Two of these maneuvers, one early in the program and another at the very end, were especially impressive and would subject Barnes to up to nine gs for just a few seconds. The two high-g stunts were, without a doubt, the toughest of the show, but he was well practiced and experienced. The rehearsals the day before and the three flights he had taken earlier that day had exacted a toll, however, and Barnes was slightly fatigued as he approached the first of the two major high-g moves.

The Tigershark streaked up and over an imaginary mountain in the sky and Barnes' body became weightless as he arched over the top of the climbing trajectory. He guided the plane into the planned negative-g maneuver, and his shoulder straps dug into his neck and upper back, keeping him from flinging out of his seat. Now he could feel the blood rush to his head and the pressure forming around his eyes. Fortunately, the routine called for no more of these negative gs, which were far more uncomfortable and much less tolerable than the normal positive gs, and he pointed her nose downward toward the bottom of the imaginary mountain on his invisible roller coaster path.

His breathing increased dramatically in anticipation of the next move, and the muscles of his arm and neck tightened as he accelerated down another invisible mountainside. At the bottom of the trough he pulled back on his hand controller and immediately felt the onset of the force - - first two, then five, then nine times the force of gravity - - all within one and a half seconds. His body was positioned perfectly within the reclined seat, and his arms and legs were all held firmly on their rests. During a maneuver like this the only thing he could move was his right wrist, which continued to pull back on the control as he swooped upward. The g-suit responded to the powerful force, squeezing hard and pushing blood out of his legs and up to his torso. The downward pressure from the gs was tremendous, and the skin on his face stretched and distorted down toward his chin. The retinas of Barnes eyes, having the highest histolic blood pressure requirement in his body, showed the first symptoms of the loss of pressure and blood flow. The minuscule blood vessels at the back of his sockets began to shrink and wither with each successive increase in g force, and his vision began to fade. Suddenly, he was looking through a donut-shaped cloud. It became darker as it expanded toward the center. Barnes had experienced this partial "gray out" numerous times before during high gs, so he was not overly alarmed. He knew his vision would return as the gs subsided and as the blood made its way back up to his eyes. The trick was to avoid going too hard or too fast. Within a few moments the massive gs released their grip on him as quickly as they had appeared, and the blood was free to flow back to his eyes and brain. His vision returned to normal as quickly as it had faded, and Barnes pulled his right-hand side controller to the left for a series of quick rolls.

Barnes was drenched in sweat from the effort required to keep his hands and feet on the controls and his head reasonably clear. That marked the seventh high-g maneuver he had completed that day and had been the most exhausting by far.

Each of the last ten seconds had taxed him to his limit, and his muscles ached from being stressed to their maximum output. Fortunately, he had only one more to go.

The Tigershark streaked toward the end of the runway traveling at 361 knots, well over 400 miles an hour. He aimed for a position southwest of the runway, just where the air show chalet and the most important observers would be located during the demonstration in Paris. The target was straight ahead and approaching rapidly, and it was nearly time for the final maneuver. He held the F-20 in a shallow angle of approach, just as if she were conducting a low-level ground attack, at the conclusion of which he would pull up and away as rapidly as possible. The gs would come on fast and strong, and Barnes' heart pounded in anticipation of the effort he would have to apply in just a few moments. Only twenty seconds to go and the show would be over.

This was it. Barnes tightened his grip around the main flight control in his right palm. All it took was a small flick of his supported wrist to send the F-20 off in any direction. He had reached the simulated target of the chalet and it was time to execute the maneuver. He pulled the control back toward him and then moved it to the right. On cue, the Tigershark raised its massive gray snout to the light streaming down from above and bolted upward, rolling slightly to the right as it reared its head skyward, its powerful tail fin helping steer the course.

The force hit Barnes hard, harder than anything he had ever felt. One second he was floating, and exactly one second later he was within the crushing grip of over six times the force of gravity. A fraction of a second later it was up over seven gs. His g-suit responded by squeezing his legs and abdomen, but not soon enough to keep the blood from succumbing to the powerful downward tug. His vision was the first thing to go. The tiny veins in his retinas collapsed as the blood in them drained out and downward. It started as a narrowing of his field of view, but then a solid gray cloud encircled the cockpit and the horizon outside. The grayness quickly billowed inward until all he could see was a small circle in front. Then the remaining circle of vision disappeared too, and David Barnes knew he was in serious trouble.

The massive downward pull continued, and again he was not fully prepared for what followed. Perhaps it was the combined effect of having done this seven times before that day or, possibly, that he was just a bit tired and thirsty. Whatever the case, Barnes felt control of the situation slipping away. He fought the crushing pressure around his chest and gasped for precious breaths of air.

The rush of blood from Barnes' head continued, and the veins in the uppermost part of his brain collapsed. By this time he could no longer see anything and his awareness of the position of the plane in the sky was dimming. He had rehearsed the maneuver so many times, however, that his inputs to the controls were reflexive, not deliberate or thoughtful actions. So the Tigershark continued her ride upward to the sky and Barnes shifted the hand controller to the left and started her in a left roll as planned. She rolled a full 360 degrees and then another 200, leaving him momentarily upside down and pointed well above the horizon. His speed had fallen to 175 knots due to the energy required to follow his upward and rolling track. Barnes pulled the controller back to the center line and held the plane in the inverted position for a full second and then again tilted the control to the left and continued the roll, all the while being blind and semiconscious from the loss of oxygen to his eyes and brain. For just a fraction of a second he remembered that he was supposed to do something important now, but he just could not recall what it was.

The nose of the F-20 dropped low, and she came through

one last leftward roll. At this moment Barnes was trained to shut off the afterburner, lower the landing gear, and begin a swooping left turn back toward the runway for the landing. But these were all things that required memory and thought, things he would not be capable of doing for another ten seconds as his brain recovered from the oxygen debt. Instead, he pulled back on his controller, then forward, and nosed the plane down toward the ground. His hand instinctively pulled back again, and she settled down into a long and shallow dive with the wings level, but just a few hundred feet above the ground. The Tigershark descended at a rate of 258 knots toward the forest ahead, where she slammed into a snow-covered hill three seconds later. The Tigershark broke up as she bounced across the windswept terrain in a thundering ball of flame, and David Barnes, Northrop Test Pilot, was killed instantly.

EPILOGUE

The Canadian Aviation Safety Board, assisted by experts from Northrop Corporation, determined that David Barnes became incapacitated from the brain-tissue hypoxia experienced during his final high-g maneuver. Fatigue due to repeated high-g maneuvers earlier in the day and Barnes' generally low blood pressure and lack of specific state-of-the-art training on anti-g breathing techniques in a human centrifuge were cited as factors in the accident.

Around the same time as the F-20 crash, five US Air Force fighters were lost after repeated high-g maneuvers, raising suspicions about the extent of the problem with new high performance aircraft. David Barnes' death, and the highly publicized loss of the F-20 prototype, the second prototype to crash, was the catalyst needed by the aviation community to recognize the pervasiveness of g-induced loss of consciousness

(G-LOC). His death increased other pilots' awareness of the G-LOC phenomenon and resulted in improved anti-g training among pilots of new generation fighters, such as the F-16, F-18, and F-20 Tigershark.

In December of 1986 Northrop Corporation terminated the F-20 Tigershark program due to lack of sales.

REFERENCES AND NOTES

Canadians investigate pilot's loss of consciousness prior to F-20A crash (1987). *Aviation Week & Space Technology,* March 23, 75-78.

Canadian safety board assesses crash of F-20A prototype (1987). *Aviation Week & Space Technology,* March 16, 89-94.

Green, R. G., Muir, H., Hames, M., Gradwell, D., and Green, R. L. (1991). *Human factors for pilots.* Aldershot, England: Avebury Technical, Academic Publishing Group.

Harding, R. M. and Bomar, J. B. (1990). Positive pressure breathing for acceleration protection and its role in prevention of inflight g-induced loss of consciousness. *Aviation, Space, and Environmental Medicine,* 61, 845-849.

Martin, T. and Schmidt, R. *Case study of the F-20 Tigershark.* Santa Monica, California: RAND Corporation.

O'Hare, D. and Roscoe, S. (1990). *Flightdeck performance.* Iowa: Iowa State University Press.

Parker, J.F. & West, V.R. (Eds.) (1973). *Bioastronautics data book.* Washington, D.C.: NASA.

Safety board urges pilot training in g-induced loss of consciousness (1987). *Aviation Week & Space Technology,* March 30, 103-104.

Second F-20 prototype crashes during Paris Air Show practice (1985). *Aviation Week & Space Technology,* May 20, 22-23.

Werchan, P. M. (1991). Physiologic bases of g-induced loss of consciousness. *Aviation, Space, and Environmental Medicine,* July, 612-614.

THE PEPPERMINT TWIST

When Cindia Cott and her father, John, entered the lobby of the trendy nightclub on 37th Street on the south side of Topeka on Saturday night, September 23, 1989, they wondered if the place would live up to its reputation as the hottest spot in town. What they really wanted was to sit and talk and have a drink and perhaps a bite to eat, and the new 1950's-theme club seemed like a fun place to go. They had heard a lot about the club. It sounded like it might be fun, something out of the ordinary.

An old rock and roll tune blasted from the hi-tech sound system as they walked toward the lounge. The song was as characteristic of that decade as the functional lines of the bar and aqua and red hues reflected in the polished chrome furniture. Why, even the waitresses were dressed for the part. So, enticed by the setting and the suggestion of a good time, they decided to stay and found a little table out on the crowded linoleum floor.

A waitress approached to greet them and take their order moments after they sat down. And, as waitresses do, she asked them politely if they wanted something to drink. They might be interested, she explained, in the special drink the club was offering that evening. It was called a watermelon shot. Tasted like watermelons, they said, and was available for only one dollar. Cindia and her father decided that a watermelon shot sounded pretty good, so they each ordered one. The price was certainly right.

The waitress worked the nearby tables and returned to the bar to place her orders with the barkeep. The watermelon shots were selling well, and she gave him an order for twelve of the sweet drinks. So, as barkeeps do, he started mixing a batch of drinks. Into a bottle he poured a couple of cups of orange juice, then an equal amount of Southern Comfort followed by a few teaspoons of creme de noyau to finish it off. He stirred it all up, and the resulting concoction turned creamy and pink, a little lighter than the inside of a watermelon. They said its taste was reminiscent of the flavorful fruit - - especially if you added a little imagination.

The batch of drinks completed, the barkeep placed the bottle and a dozen shot glasses on the counter for them to be picked up by the waitress when she returned from her rounds. She was back to the bar not a minute later and retrieved her clean glasses and drinks and walked toward the tables balancing a tray holding the shot glasses and a bottle of creamy pink liquid.

But the bottle on her tray did not contain the watermelon-pink mixture of orange juice, Southern Comfort, and creme de noyau. Instead, it held a fruit-colored brew of corrosive chemicals identical in appearance to the watermelon beverage the barkeep had just blended. The attractive mixture's main active ingredient was not orange juice or Southern Comfort. It was sodium hydroxide, more commonly known as lye, and a group of unsuspecting patrons in The Peppermint Twist nightclub were about to get the shock of their lives.

Back behind the rear wall of the bar, back where they cooked the food and cleaned the dishes at the club, was an industrial-strength dishwashing machine. This machine was special, not so much because it was so big and rugged, but because it was designed to operate with a unique kind of dish soap. Like most automatic dishwashers, it sprayed jets of water mixed with grease-cutting additives onto the soiled plates and tablewear, leaving them clean and sterile for the next meal. But unlike the common residential dishwasher, this industrial-strength machine used an industrial-strength dishwashing fluid known as *Eco-Klene*. Its primary active ingredient was highly caustic sodium hydroxide.

Eco-Klene, manufactured by Ecolab of St. Paul, Minnesota, was developed specifically for use in "closed" dishwashing systems such as The Peppermint Twist's industrial dishwasher. It was delivered periodically to the nightclub in semi-transparent 5-gallon plastic containers. Ecolab, well aware that their product was hazardous, described the dishwasher soap and warned of the hazards associated with its use on a label on the side of the bucket. There were no poison labels or symbols, but there were warnings spelled out in the text. The fluid could cause chemical burns and blindness, the label said, and was to be handled only while wearing gloves and goggles. A telephone hotline number was provided in the event of an accidental exposure. The company had also sent cautionary literature to their customers, including The Peppermint Twist. But on that Saturday night in September, the supplemental literature about the use of the product was, as one might expect, tucked away in a file cabinet.

The text on the side of the plastic bucket was printed in red ink, a seemingly appropriate color for a warning label and one which looked attractive considering the design of the total package. But like many household products that come in those shapely and attractive plastic squeeze bottles, the creamy pink cleaner showed through the sides of the bucket, giving the container a pleasing pink color. The shaded pink plastic made the red text less obvious than it might have been had the warnings been more prominent or printed on a higher contrast background.

Replenishing the machine with *Eco-Klene* was simple enough. When the dishwasher ran out of fluid, a full 5-gallon bucket was carried over to the machine and the spout removed. The *Eco-Klene* was automatically drawn into the machine during each wash through a hose inserted into the spout. The fluid was not really meant to be poured from the container, but pouring fluid from the bucket was easy enough. All one really had to do was lift it up and tip it over a bit. There was not a simple check valve or similar feature to prohibit pouring through the opening. And there was little indication, beyond the information provided on the label and the company's literature, that the dishwashing fluid was to be used exclusively in conjunction with the dishwasher and no where else. Ecolab representatives made periodic deliveries of the containers and offered orientation training, but, as witnessed by some kitchen employees, they did not always wear gloves and goggles when handling the material, perhaps contributing to the impression that the fluid was not all that toxic.

A few weeks before that Saturday night in September, the employees working at the bar ran out of regular liquid dish soap. They requested that some *Joy* or other common dishwashing liquid be purchased, but somehow the order fell through the cracks, and the dish soap never arrived. Faced with piles of dirty glasses in the two sinks, the barkeep went to the kitchen, pulled out one of the pink 5-gallon buckets of *Eco-Klene*, and poured some out of the spout into a bottle. It was one of the bottles that they used to serve mixed drinks. He returned to the bar with the fluid and used it to wash the dishes. The barkeep didn't think of reading the label on the 5-gallon bucket, and he had not seen any of the warning literature from the manufacturer. "Soap is soap," he would say later.

So for a number of weeks The Peppermint Twist employees working at the bar washed the glasses with the *Eco-Klene* dishwashing fluid in the two sinks. And on that fateful night, someone chanced to set the open bottle of *Eco-Klene* upon a small cabinet near the counter. The bottle was similar to the one the barkeep used to mix the watermelon shots, and the *Eco-Klene* was visually indistinguishable from the nearby watermelon-flavored mixed drink. It did not have a strong odor, certainly not a smell suggestive of caustic cleaning fluid.

The waitress approached the table, delicately set a napkin down, and placed a shot glass full of the watermelon-colored liquid in front of Cindia. Cindia's father got one too. The waitress walked off to distribute the remaining drinks, then headed back toward the bar.

The shot glasses were raised in a silent toast. And, as bar patrons do, Cindia and her father each took a full gulp of the contents in the small glass. The reaction was immediate. The caustic cleaning fluid started dissolving the lining of her mouth, throat, and esophagus, burning and destroying the living tissues. She clutched her throat in a deathly grip. God. The pain. It was obviously the drink, but what was it? It was excruciating, especially in her chest, and it was getting worse! It slid down her throat and deep into her chest and then back up to her head like a flame ascending the sides of a post of dry wood.

A growing commotion erupted around her, for others had also swallowed the drink. But Cindia was focused on her own perilous situation. She was helpless, and she didn't know what to do. Then the air wasn't getting to her lungs. She started gasping for breath, and she feared that she was going to die.

Word spread quickly back to the bar that there was something wrong with the drinks. People out at the tables were violently ill. "What did you do, what did you do?" they said to the waitress. All she had done, of course, was serve the customers their drinks. Well, "return the customers' money and keep quiet," she was told by the employees back at the bar.

Within minutes it was determined by the cook that the dozen customers had been served the caustic *Eco-Klene*. Someone called for the ambulances, and the patrons were rushed off to the hospital emergency room. Cindia Cott, perhaps the most severely injured of the group, had severe burns in her esophagus and the muscles used in swallowing. Her weight fell from a petite 107 pounds to 75 pounds in only 11 days in the hospital. There she learned that her esophagus was seriously damaged, and it would continue to constrict and grow scar tissue, perhaps for the rest of her life. She could look forward to undergoing a monthly throat dilation procedure or have her damaged esophagus replaced with a section of her own colon.

EPILOGUE

Two years later, after The Peppermint Twist closed its doors and went out of business, her legal case against the club came to trial. She had previously reached an undisclosed agreement with the manufacturer of *Eco-Klene*, Ecolab. One witness for the nightclub was a former bartender from Seattle who had taken a gulp of *Eco-Klene*, mistaking it for a cup of his regular drink of cranberry juice. The *Eco-Klene* had been brought to the bar in a cup and was being used to wash dishes. His injuries were extensive and permanently debilitating.

Although not discussed during the trial, it came to light that five elderly hospital patients in British Columbia were fed a powerful industrial-quality dish soap called *Mikro-quat* in September of 1986. The liquid had been transferred from a 217-gallon container into an unlabeled 4.5-gallon container. A dietary aid poured the "pink fruit juice" into cups and distributed it to the patrons just before bed time. Like the *Eco-Klene*, it was an attractive pink color and did not have a noticeable or unpleasant smell.

On September 26, 1991, a jury awarded Cindia Cott, her father, and another nightclub patron 3.3 million dollars in damages.

REFERENCES AND NOTES

Doctors testify in nightclub case (1991). *The Capital-Journal*, September 19, 4D.

Ex-nightclub employees testify (1991). *The Capital-Journal*, September 12, metro section.

Inspector couldn't tell shots from soap (1991). *The Capital-Journal*, September 13, 2D.

Judge rules in poisoning (1991). *The Capital-Journal*, September 20, 9E.

Peppermint Twist customer thought she was going to die (1991). *The Capital-Journal*, September 14, 2B.

Physicians testify at poisoning trial (1991). *The Capital-Journal*, September 17, 2D.

Poison case testimony near end (1991). *The Capital-Journal*, September 25, 10A.

Seattle man "terrified" after his poisoning (1991). *The Capital-Journal*, September 18, 1A.

Suit against nightclub for poisoning goes to trial (1991). *The Capital-Journal*, September 11, 1A.

Soap's hazards never described, manager testifies (1991). *The Capital-Journal*, September 21, 2B.

Trial coincides with anniversary (1991). *The Capital-Journal*, September 24, 2D.

Twist case goes to jury (1991). *The Capital-Journal*, September 26, 2D.

Watermelon shot not too popular in Topeka (1991). *The Capital-Journal*, September 13, 2D.

3 awarded $3.3 million in damages in Twist trial (1991). *The Capital-Journal*, September 27, 1A.

5 elderly patients recovering after disinfectant mistaken for juice (1986). *The Herald*, September 11, 16C.

THE PRICE OF THE AMAGASAKI

There could be no sound and no splash and nothing for other eyes to see that night as Royal Navy Lieutenant James Kull slipped into the water from the boat and kicked submerged across the wide channel to the Japanese ship at anchor 2,000 yards away. "Become a crocodile," his Aussie instructor liked to say. "Every ripple of water, every clank of metal on metal, every reflected sparkle of light will alert your prey. It is to be his life or your life. There is no in-between, and the outcome is yours alone to make. You must remain undetected. Attack with stealth."

Stealth was on Kull's mind as he sat in the rubber suit on the gunwale of the gray rowboat with his legs inside and fins flat on the floor. It was past the time to go. He looked up and beyond his clipped nose, through the round plate of the rubber mask, at his accomplice at the oars. Kull touched the cock valve on the mouthpiece at the end of the corrugated rubber hose that ran in a vertical arc from his sternum to his mouth. The valve to the outside was closed. Through the rubber mouthpiece clinched between his lips and teeth he inhaled the pure oxygen contents of the counterlung in one breath, then he pressed down on the valve to send his exhaled breath of nitrogen and carbon dioxide and oxygen out into the night. He released the valve to suck in only the pure oxygen from the rubber counterlung and the small bottle of pure oxygen that fed it. For two long minutes by the luminous dials of the watch on his left wrist he carefully repeated the cycle, each time filling his chest with the pure oxygen and then exhaling to the air outside through the valve on the mouthpiece.

He let go of the valve for good. The artificial lung strapped to his chest deflated synchronously with the expansion of his own inflating lungs as the gas flowed through the rubber hose out of one and into the other. He breathed out fully and the bag on his chest inflated, and it deflated again as he breathed back in. It would be this way until daybreak many hours away, a pendulum exchange of gas from the artificial lung into the real lung and from the real lung to the artificial lung. It was now, as the engineers liked to call it, a closed loop.

Kull lifted the 30-pound bag from the floor and placed it in the water where it sank slowly below the surface. The bag was attached securely to his waist with an eight-foot line. It was now taut and he knew that the bag had sunk to the side underneath him where it needed to be. He spread his open right hand around the perimeter of the round face plate of his mask, carefully shifted his balance, and slithered smoothly backwards off the side and into the calm water of the Johore Strait between the island of Singapore and the mainland of

Malaysia to the north. It was deep into the night of March 10, 1944 and one thousand miles deep into enemy-held waters, and the last phase of his mission had begun, a mission to singlehandedly, in the end, sink one of the mightiest ships of the Imperial Japanese Navy, the cruiser Amagasaki at anchor just over a mile away under a star-lit sky.

Three limpet mines, each a thick charge the diameter of a dinner plate, were packed carefully inside the hydrodynamic bag at the end of the line attached to the canvas belt on his waist. His gear was heavy, but constructed and packaged to be of slight negative buoyancy when in the sea. Now Kull was weightless as well, suspended a few feet under the surface of the water that was lighter and clearer than it had been from above. It felt cool and it was good to finally be in the water.

He lay suspended on his back beneath the surface, let the little pockets of air in his equipment and fins and head-to-ankle rubber suit escape, and watched the small crystal globes against the dark but clear marine backdrop of the underwater sea at night float up to the iridescent surface above. He drew a few calm and slow breaths from his mouthpiece while the last bubbles cleared his gear and made their way upward next to the underside of the boat.

The rebreather was working well. The oxygen seemed pure, only slightly tainted by the rubber hose and gas bags of the closed-circuit system. There were now no bubbles in the water at all. There was just an endless and almost mystical supply of easy-breathing pure oxygen from the rebreather. There was no noise and there were no bubbles from his exhaled breath which flowed from his own lungs into the external lung and then back again.

He had now drifted under the boat in the direction of the Amagasaki. He gave a long and easy kick with his fins and ascended slowly, his line and bag of limpet mines in tow, all the while careful to breathe out as he went up. Kull slowed and silently broke the surface with his back toward the Japanese cruiser well off in the distance and his reflective face mask toward the grease-painted face of his fellow saboteur still in the boat. Kull gave a thumbs-up with his hand in front of his mask and the gesture was returned by his accomplice hunched low to lessen his profile on the horizon. The diver turned to the distant Japanese ship and checked its bearing with the compass on his left wrist. The drop had been right on target and there had been no ripples and no reflections of light to alert his prey. He stopped his long easy kicks and let himself sink as slowly and as silently as a crocodile beneath the surface.

His rate of descent increased as he passed through 10 feet and he took a shallow breath to expel the carbon dioxide and replenish the oxygen in his blood. Many days and nights had been spent calculating and testing his buoyancy in the preceding week in the event that an important ship presented itself in the harbor. Everything was right. He was slightly negative after he sank below 15 feet and he knew he would be even more negative the deeper he went, but not enough that he would have to struggle to stay deep or to ascend. His forward motion and prone position would make it easy to go up or down at will once he was underway.

Kull and the members of his team had planned every minute of the mission and every kick of the fins on his feet. He would not swim directly towards the ship, but take three connected straight legs, each for a predetermined time at a calculated speed over a known distance. Once between the first and second leg, and then again quickly between the second leg and the final long third leg, he would rise to just below the surface, face the ship based on his compass reading, gently rise above the surface long enough to get a fix on the ship, and sink quietly below for a look at his compass to check and adjust the next leg of his underwater swim. They had studied and planned around the movements of the tides and the moon, the known currents in the strait, the silhouette of the shoreline, and the lines of visibility from the ship, all to maintain stealth, all to accomplish the mission. Far too much was at stake to have done anything less.

The light that night was brighter than they had expected, given the dark phase of the moon and time of year here along the equator. But the humidity was uncharacteristically low and the stars reflected just enough light to see the shore, a ship, and even an underwater swimmer if he drew enough attention to himself. Lieutenant Kull had the ability and skill to swim for miles at depth in water as black as ink and set the fuse on a mine on the hull of a ship in the worst of swift and muddy currents, but tonight the water, like the air, was exceptionally clear and calm. Maintaining stealth would require that he maintain depth, something he should not have trouble doing courtesy of his closed-circuit rebreather.

As he sank slowly on his way to a depth of 30 feet with the sack packed with three limpet mines hanging down lightly below his fins, he watched the depth gage strapped to his wrist. At 20 feet he leveled out and began to kick with full and experienced strokes of his long rubber fins. His body flattened out in the water and the line and attached sack of mines trailed behind as he moved on. He reached back with both hands and rotated the canvas belt about his waist a few inches so that the tow line streamed back nicely and comfortably from his belt and then between his legs and fins. There was over a full nautical mile to cover underwater, and, although it was heavy and awkward on the boat, the bag could be pulled along easily as long as the line was trailing back so that he did not kick it.

Kull raised his left arm with his elbow bent at 90 degrees and placed his wrist in front of his mask to see the luminous dials of his depth gauge, watch, and compass. The compass face was now horizontal and the round dial beneath the glass rotated to point north. With his right hand he rotated the round bezel on the compass to mark the bearing of the first leg of his swim, twisted his shoulders slightly to adjust his heading, and swam off along his course. Kull grasped his left fist with his right hand and assumed the posture he would hold for most of his long swim to the Amagasaki. The forearm, with his depth gauge, watch, and compass, would be a panel of florescent instruments, just as in a dark cockpit flying through fog or over the ocean on a moonless night. Other than the occasional shimmer of light that walked its way down through the water when the ripples on the surface aligned by chance with the light from a group of particularly bright stars, all around him was dark and calm. Kull raised his bent left arm slightly higher to get as streamlined as possible, tucked his right elbow close to the side of his chest while still grasping the fist of his left hand, and lowered his head to reduce his resistance. The luminous second hand on the watch swung past 12 and he paced the rhythm of his kicks and counted their number relative to the sweep of the dial. He would swim along the bearing for only eight minutes and kick at the pace of 100 beats per minute while maintaining the length of each stroke of his fins from the top of the kick to the bottom of the kick. A constant and known bearing and a known and constant speed at a constant and known depth would put him at a predetermined point well clear of an underwater outcropping and in position for his next leg in precisely eight minutes.

The rebreather that he wore was a Jules Vern-inspired blend of black organic tucks and rubber atop geometric shapes that can be made only by a machine. A rubber hood sheathed Kull's neck and head and face all the way around the cyclops face plate of his mask. Each exhaled breath flowed out his mouthpiece, down a thick rubber hose, and into the center dome of a round chest-mounted canister that Kull had packed with granules of soda-lime to absorb the carbon dioxide of his exhaled breath as it flowed into the rubber gas bag inside his rubber and canvas vest. For each volume of carbon dioxide absorbed from his breath, an equivalent volume of pure oxygen was bled into the rebreathing bag from the small pressurized tank. Only carbon dioxide from his lungs was added to the pure oxygen as he exhaled, and only pure oxygen was added to the contents of the rebreather bag to replace the volume of carbon dioxide extracted by the can of CO_2-absorbing soda-lime. He could remain submerged for hours, emitting no bubbles and no sound except for the occasional squeak of rubber on rubber from his fins and gear or his teeth readjusting their grip on the mouthpiece.

Through dark liquid space he kicked and towed the precious payload of limpet mines behind on the tether. The three luminous dials on his left forearm covered the round field of view through his mask, and he concentrated on keeping his left elbow up and pointed forward and his right elbow tucked in close to his chest to lessen the drag in the water. His breathing was now heavy and rhythmic and in synchrony with the beating of his heart and the timed pace of the strong kicks of his legs and fins.

Keeping a straight course was easy as long as he kept the compass flat and its dials from rubbing against and hanging up on the underside of the glass. The watch took care of itself regardless of the slant of his forearm, but the pace of work was demanding and Kull concentrated on his rhythm, the sweep of the second hand, and the elapsed time since he had gotten underway.

The depth of the water the length of his planned route as well as where the Amagasaki lay at anchor was 60 feet, and it came as an abrupt shock when his arms and mask slammed against the sandy bottom and the silt stirred up in front of the dials of his instruments. He had gone too deep, much too deep, and Kull quickly read the depth on the gage. It was 58 feet, nearly three atmospheres of pressure. He never intended to drop below 30 feet on the first leg, but had concentrated so much on keeping the compass flat to get an accurate bearing that he was not aware of his slowly increasing depth. Once above and clear of the stirred-up silt, Kull looked upward and could just discern the surface where air met water, something he had rarely seen from this depth at night. Fluid ripples caught faint points in the sky and on the horizon toward Singapore and reflected them downward. He mentally added thirty seconds to his required swimming time for that leg and got under way again, angling upward but along his bearing until he reached his planned depth of 30 feet.

Minutes later he stopped the long rhythmic strokes of his fins and hung suspended at an equal distance between the surface and the bottom. It was the hopeful end of his first leg and the spot for his planned sneak and peek. His breathing was labored and he suddenly felt nauseous, not knowing if it was from the extreme exertion, the demanding pace, or something else. Kull kicked slowly upward, but it was difficult to tell which way was up. The water was not as clear as in the area he had crossed and he could not see the surface, the bottom, or anything else. The bag of limpet mines was slightly negative, so the tether attached to his belt must be hanging straight toward the bottom, and he felt it with his right hand and knew that he could follow the imaginary line in the water straight up to the surface. The depth gauge readings decreased in step with his kicks; he was going up after all. At five feet he paused and turned toward the bearing at which the Amagasaki should be from this mark. Just below the surface he hung suspended again and turned his face precisely toward the memorized compass bearing. This first leg was to have taken him out and beyond an outcropping from shore and placed him in a good position for the next step. If his navigation had been correct he would see only the stern of the ship in the distance as if she were sailing away from him down the wide channel.

At the surface his mask would stay on, his nose clipped, the mouthpiece held tightly between his teeth and lips, and the cock valve closed. Should he take a single breath from the outside and then resume breathing with the rebreather he would quickly suffocate. It all seemed so illogical - - that breathing pure air would kill him - - but the nitrogen in the air, once introduced to his closed system, would never be absorbed by his lungs. Back underwater, he would take it in and breathe it back out. The total volume would never decrease - - as it did when he burned up the pure oxygen - - and the rebreather would never add more oxygen from the pressurized bottle into the counterlung if the bag were never completely empty when he drew in a full breath.

The half circle of his face plate raised above the surface to his eyes. Directly ahead, although still a good distance away, sat the hulking stern of the Amagasaki. He slipped back below the surface just as quickly and smoothly as he had risen above it, careful not to change his orientation toward the ship. Below the water Kull checked his compass one more time. He was right on the mark.

As he sank back to depth, Kull twisted the compass bezel to set the new heading and set the bezel on his watch to time the duration of the next leg. Straight down he sank, relaxing more or less for a moment and recovering from the exertion of the past minutes. At 40 feet he arrested his descent and headed off horizontally again on his new course with his rhythmic and counted kick and luminous instruments squarely in front of his face.

War calls for extremes, sometimes the ultimate extreme, and the mission and the technology that enabled him to carry it out with stealth was extreme in the strictest sense of the word. That a man could breathe underwater from air pumped down to him from the surface or even breathe from a tank or container of air brought down with him was one thing; it was another matter entirely to think that a man could swim underwater by breathing out into a bag and then breathe its contents back in after the exhaled carbon dioxide had been removed. He would, of course, eventually burn up the oxygen in the bag and it would empty, collapsed from having its contents sucked out, used up, exhaled, and packed away in a little can of chemicals. But more oxygen would flow to the bag from the small tank carried with him. Kull's compatriots in Britain had pioneered the technology during the war, and, among the Allies, the Royal Navy was the staunchest proponent of oxygen rebreathers due to their nearly silent, non magnetic, and bubbleless operation. But the technology had been introduced rapidly due to the need, and there had been a few mishaps along the way.

There was, for reasons not explained, the occasional diver who developed spasms, fainted, or convulsed after using the system at depth. The Navy thought it might be due to breathing pure oxygen under pressure, and the deeper one went the more pressure there was. In view of the occasional accident and suspicions that they might not know everything they might need to know, Kull had orders not to exceed a depth of 15 meters, or slightly more than 49 feet, about two and a half times the atmospheric pressure at the surface. Staying above this relatively modest depth should keep him out of trouble, it was reasoned, and it was certainly not as if the gas in the rebreather was toxic or poisonous. He was breathing pure oxygen after all, the stuff of life.

The current moved faster now but as if from upstream, from the waypoint where he needed to be next. There had been no current at the surface or on the way drifting back down, but now it was quite strong at 45 feet. The air hose from his mouth to his chest vibrated and made a slight noise for a second and then relaxed as each fin swung in an arc on the downbeat of each kick. To keep to the schedule he had to pick up the rate, and Kull cranked the pace up a notch and counted his kicks as the second hand swung around the center dial on his wrist.

This was another short leg. The work was hard and he sweat heavily under the rubber suit. The perspiration did not evaporate or wash away to cool his muscles and his blood. The sweat rolled down his brow beneath his mask and into his eyes where it stung, and he blinked and closed his eye lids tight and rolled his eyes to help dilute the salty sweat while kicking and holding the course both left and right and up and down, as the seconds ticked by and as his outstretched legs pumped hard.

It started on the left side of his upper lip, as if a fishing line were in the muscle, pulling and twitching from an unseen presence on the other end. Kull loosened the grip of his teeth around the mouthpiece and moved the muscles of his lip and stretched them out to lengthen the muscle and nerve. It seemed to help, but then it happened again, only harder this time and from higher up in his cheek. It stopped as quickly as it came on, but seconds later it was back; he released his grip on his left fist with his right hand and felt his cheek and it was taut and tight and his lip was pulled up and twitching as if hooked by a line. He rubbed it with his hand and it felt strange because nothing like this had happened before during his dives.

He kicked on at 45 feet along the bearing while watching the round dial of the watch and waiting for the minute hand to reach the mark on the bezel to tell him that the leg was over and to check his position. With each minute his breathing ran faster, but also shallower and more labored.

On one inhalation he first noticed the pain in his chest under his sternum below the place where the air hose attached to the CO_2 scrubber at the center of the vest. The pain was deep in his chest and it started out small and then grew with each inhalation. He broke the rhythmic pace of his breathing and his kicks and coughed to clear his throat which had become dry and painful, but his throat would not clear and the pain under his sternum did not subside. Instead, the pain spread and clutched his windpipe below his throat and made him cough again to clear the obstruction and moisten the lining. Kull kicked onward through the black ink, redirecting his focus to the pace and the bearing and the depth until the luminous minute hand in front of his face reached the tick mark.

He was finally there, or so he hoped, and the tether with the mines at the end sank below him when he stopped and pointed the way upward to the surface from 45 feet down. It was not a problem to see which way was up now, as the water was dark but crystal clear and the light danced on the surface as Kull kicked slowly upward. At 15 feet he heard a clank, and then another, and turned his head left and right to localize the source of the sound in the water. He held his depth with a constant easy kick and waited in silence for another sound. It came again, and he heard the faint whirr of a blower that faded and then grew louder as the sound traveled through the moving currents and eddies in the strait. He estimated the bearing to the sound with his compass and it matched the memorized direction for the final leg of the swim.

It was the Amagasaki, the sound of a closing hatch here or there, the grinding of her anchor chain, the hum of a fan. She was sitting broadside to him, and if he only swam on the bearing from this point and followed the sound he could not miss her. And the water was so clear and the sky so bright now even without the moon that he stopped his ascent just under the surface, decided not to risk being seen, and simply sank back down to depth and started in on the new bearing to the target. There was no question where he was and no need to take the chance that a pair of eyes from the ship might happen to glance his way just as his mask cracked the surface and reflected a stray bit of light back to the ship. He would just go in, aiming slightly to the right, where the bow should be pointed up into the current.

Back at depth just shy of 50 feet the water was even clearer than before. The bottom was sandy and clean and reflected the dull backdrop of an ambient glow in the sea that only an adjusted eye could detect. Kull felt the need to stay deep, right up to the 15 meter limit that he had been given. This was the longest of his three legs, and he lined up his bearing and settled into the synchronized pattern of breathing and kicking, not concentrating so much on the pace and the clock now that he knew the target lay ahead and there would be no more waypoints. There were other things than his speed to worry about. The pain in his chest had returned with a vengeance, and the gnawing in his throat and chest had gone past the point of coughing or grunting voluntarily to clear things out. The discomfort was boiling now, out of control, and the reflex to clear his lungs and chest took over. A muffled burst of harsh barks echoed through the rebreather and into the water when he could no longer suppress his brain's command to cough and cough again. The mouthpiece stayed fixed between his teeth and the equipment remained watertight when he coughed the pure oxygen up through his chest, out his mouth, and into the counterlung; the worry was that he might somehow be heard on the ship just as he could now hear them.

The ship grew closer and the sounds from the Amagasaki drew nearer, but not before the nerves in his face fired off again, this time with more force, and his lip and cheek pulled and twitched violently beneath the rubber hood and mask. The coughing spasms came fast and strong, and Kull dropped his arms straight along his sides to streamline his form to reach the ship with what energy and control he had left. He would swim for the sound. She had a big deep draft, and he would run straight into her hull as long as he followed the trail of noise and did not go all the way under her by mistake or break the surface and be seen.

Although he had known about the winds and the tides, navigating underwater, the preparation and balance of his equipment, and how to place a mine on the hull of a warship in the dark of night, it was what neither Kull nor the Royal Navy knew in the early Spring of 1944 that now posed the greatest danger of all. It was the pure oxygen contained in the little tank that fit snugly in his pack atop his abdomen. With each breath that he had taken during the preceding hour, Lieutenant James Kull of the Royal Navy had been poisoning himself. At sea level and in normal air, oxygen comprised 21 percent of what he breathed. With 100 percent oxygen and the added weight of the water, which piled atop another full atmosphere of pressure with each 33 feet of depth, the most basic physiological processes of his body were altered. The hemoglobin and the liquid in his blood were now saturated with pure oxygen. Rather than consume the oxygen in his hemoglobin, his tissues used the excessive dissolved oxygen in his blood. The waste carbon dioxide was not carried away by the oxygen-saturated hemoglobin. One after another of the hundreds of chemical processes within his cells broke down due to the gas imbalances. Cellular metabolism slowed, enzymes stopped working their tricks, and the transmission of impulses in his nerves was increasingly impaired. Kull's veins and arteries constricted further with each breath. Less blood flowed to his organs, his swelling heart pounded away, and his thoughts became more confused. It had started with his body activating its defenses to the onslaught of oxygen, fending off the attack by coughing it out and warning with pain, but it had cascaded into an avalanche of physiological collapse in which one organ's actions or inactions were affecting everything else with the growing certainty of fatal outcome if something did not change soon.

But the water was so clear and the ship so close that he felt certain of his stealth only by staying deep. And deep he stayed, kicking with his cargo in tow until another coughing spasm forced him to stop for a few seconds when the pace of his swimming was broken by the explosive constrictions of his chest and lungs. Despite the pain in his chest, his cough and dry throat, Kull did not know that it was oxygen toxicity that was to blame or that chemical and physiological reactions were ravaging his body. That he had never shown any symptoms before on any of his other dives did not matter, just as it did not matter now that he had been a fit and very determined man only an hour before. Oxygen poisoning could take out one man and not the other, and the man not attacked might convulse and die without any warning on his next dive to a shallower depth a week later. It had nothing to do with youth or swimming skill or much of anything else, except for one thing, and that was breathing pure oxygen at pressure and the mass of random things that make humans human. And the Navy had gotten the pressure limits all wrong, not by malice or carelessness, but by ignorance and the calls of war and the availability of a technology that provided unparalleled stealth.

The reasonable safe limit of depth for a pure oxygen rebreather was not the 2.5 atmospheres or 49 feet that the Navy had dictated, but less than half that much, and each time a diver surpassed very shallow depth with pure oxygen he risked death by oxygen toxicity, a fact that would not be well understood until after the war when there was time for studies and diving physiology was much better understood.

A metallic thud came from off to his left and another sound, an odd groan of metal sliding against metal, sounded from forward and to the right. She was out there, long and hulking and broadside straight ahead. The water was of a slightly lighter shade, but only halfway to the bottom, and Kull saw his target finally after all this time, the gray steel wall of metal that ran across as far as he could see, floating above an even darker bottom that the massive ship held in faint shadow. She was huge, and the sight of his target refocused his attention away from his agony and back to his mission.

Kull turned off to the right and kicked up to intersect her closer to the bow. He came in under the starboard side and rolled over onto his back once under the ship and rose up and touched her keel with his bare hands. There were no bubbles venting from his rebreather to float up to the steel and then bounce out along the rolling bottom along the sides to the surface to be seen. And there was no sound, other than the little clicks and pops one hears in the ocean at night and the occasional noises of the ship from the other side of her steel hull. He had made it.

Rolling back over, Kull cruised along her keel slightly more toward the bow and into the smooth current. There was just enough light to see the steel ceiling inches above his head. The spot was good, and he stopped and pulled up the tether on his waist belt, loosened the tie on the bag without trying to look at what he could not see, and pulled out a limpet mine from its secure pouch inside. The bag settled back down and dangled beneath him, and Kull kicked off to the side a few feet holding the mine like a discus in his hand against his right breast. Face up again, he found a place away from the strong solid keel and between strong ribs where her belly would be softer and the charge could blow a hole through her thick steel plate. The magnetic back of the limpet clicked onto the hull. Kull felt the dial to find the top and to count the holes and stuck his finger in the correct one and rotated the dial to the stop. He pushed the pin into position and the mine was armed and set to blow in 60 minutes.

Pain flowed in thick waves through his head, through the constricted and deprived veins of his brain as he flipped back over and kicked off down the keel in the direction of the stern. But everything must be perfect now. There was no room for distraction or error. The light current pushed him along under the ship, and in a hundred feet he stopped to quickly pull the bag up again. The tether seemed to have looped around his leg, and as he brought down his chin and pulled up his knees to move the line from around his foot or his foot around the line the pain welled up in his chest even worse than before. It balled up in his throat and he coughed into his counterlung as it all came back in a flood of pain. He had to push past it. There was no stopping and no turning around. Kull choked and hacked again but caught a breath of oxygen and held it in as he removed another limpet from the bag and felt for a smooth sheet of metal away from rivets and underlying steel up underneath the ship. It latched onto the steel hull with a solid metal snap and he found the right finger hole and rotated the dial to set the time. He armed the mine and reached for the bag for the third time.

It had to be here and now. The attack had been staged from upcurrent to down so he could swim fast from one placement of a limpet to the next, but also to use the current if there were a problem. The current carried Kull adrift beneath the ship another 40 feet while he struggled with the mine and held it ready to stick on. His face up to the sloping bottom again, he ran his shaking hand over the steel plate to find a smooth spot off to the side just away from the keel. Up came the limpet mine from his cramping hands to the hull, and the magnet stuck to the steel. The fire in his chest moved on to his arms, and he kicked up hard to press against the ship and brace his spread elbows against the hull. With the telephone dial timer an inch away from his mask, he counted the holes with the tip of his finger and rotated the dial to the stop.

The seizure was well underway when he shifted the pin to set the charge. The job done, he let go and his jaw clinched down and his teeth cracked under the uneven vice-clamp of his bite on the mouthpiece. The neuroelectric charge pulsed through his brain and body, and he grunted in desperate triumph in his last moment of full consciousness as his arms bolted straight out to the side and his legs locked solid in a wooden catatonic seizure brought on by oxygen toxicity. He began to sink, slowly at first and then faster with depth, frozen in his scarecrow stance, a block of cement falling through space to the sandy bottom beneath the mighty Amagasaki.

He awoke long afterward, hazy and confused and remembering nothing of where he was or how he got there. Thirty feet straight up above hung the bottom of a solid long mass that displaced the crystalline underside surface where the air meets the sea at night. Light danced here and there around the perimeter at the top and walked its way down when the ripples happened by chance to align with the brightest star in the morning sky. A pair of upturned glassed-over eyes, still perceiving and still alive beneath the night shadow of the ship, saw the three successive flashes of white light directly overhead and the untold tons of steel fill with water and settle straight down to the sandy bottom of the Johore Strait deep into the night of March 10, 1944. Such was the price of the Amagasaki.

REFERENCES AND NOTES

Adolfson, J. and Berghage, T. (1974). *Perception and Performance Under Water*. New York: John Wiley & Sons.

Balentine, J. D. (1982). *Pathology of Oxygen Toxicity*. New York: Academic Press.

Bennett, R. B. and Elliott, D. H. (1982). *The Physiology and Medicine of Diving* (Third Edition). London: Bailliere Tindall.

Connell, B. (1960). *Return of the Tiger*. New York: Doubleday & Company.

Cushman, L. (1969). Cryogenic rebreather; first public unveiling of Sub-Marine Systems and their unit. *Skin Diver*, June, 29-87.

Donald, K. W. (1947). Oxygen poisoning in man. *British Medical Journal*, 1, 172.

Mekjavic, I. B., Banister, E. W., and Morrison, J. B. (1988). *Environmental Ergonomics: Sustaining Human Performance in Harsh Environments*. New York: Taylor & Francis.

Miles, S. (1964). One hundred sixty-five diving accidents. *Journal of the Royal Navy Medical Service*, 50, 129.

Miles, S. (1969). *Underwater Medicine* (Third Edition). London: Staples Press.

Poulton, E. C. (1972). *Environment and Human Efficiency*. Springfield, Illinois: Charles C. Thomas, Publisher.

Thompson, W.A.R. (1935). The physiology of deep-sea diving. *British Medical Journal*, 2, 208.

This story of the mining of the Amagasaki by Royal Navy Lieutenant James Kull is based on a description by Cushman (1969).

MURPHY'S LAW AND NEWTON'S LAW

Rocket Technician Anders Svanson and his colleagues in the elongated assembly hall at the European space range cautiously stepped through their checklists on their clipboards, careful to not miss a thing, making certain that every part and every setting was exactly as it should be. It was the team's job to complete the work today for tomorrow's launch above the arctic. The Orion rocket on the horizontal stand in the building would be shot nearly straight up into space from the pad after final testing of the systems. The payload was secure, its circuits and parts in order, and the other important components had all checked out according to the test specs. Anders had found it necessary to tighten down a screw here and there and add an extra wrap of tape around a small bundle of wires, but the rocket was in first-class shape overall. In fact, everything seemed to be perfect. There was no reason at all to think that their string of 350 consecutive successful launches here at ESRANGE in northern-most Sweden was about to come to an abrupt and tragic end on this day, Saturday, February 27, 1993.

Tomorrow, the Orion rocket would carry a German payload into space on a parabolic trajectory up through and out of the atmosphere and then back down to earth. The instruments would measure the ozone layer in a band 20 to 25 kilometers high in the stratosphere. ESRANGE was north of the arctic circle and the ideal launch point for a study of the ozone layer at the top of the world.

Ozone - - or the lack of it - - was certainly a hot topic these days, especially after NASA, the American space agency, published those compelling images of the so-called ozone hole over (or was it underneath?) Antarctica. Anders had heard the nay-sayers, those who maintained that ozone depletion over the poles was nothing more than a normal occurrence due to natural climatic variation and the occasional volcanic eruption. But the theories seemed to make sense and, so far, the measures and observations from space appeared to be confirming the theories. And there was indisputable evidence that the atmospheric emissions from Eurasian industry were working their way up to the arctic, creating a measurable layer of smog over the north pole. It was of potential importance to everyone, but especially those in the northern latitudes who would be most affected by ozone depletion. The German government felt that the topic was certainly worthy of study and wanted their scientific payload lofted into space above the arctic on a sounding rocket to sample both atmospheric chemistry and the ozone layer.

Anders Svanson went about his business as he always had, working in concert with his Swedish associates of the Swedish

Space Company at ESRANGE. They also had the help of a German technician closely involved with the payload of scientific instruments. The Swedes and the German coordinated their activities as they huddled over the rocket lying on the test stand at chest height. One technician often talked through the test sequence on the printed forms, another attached wires for tests, while a third operated a meter or test kit. To the casual observer the process might have looked surprisingly informal given their casual attire and academic demeanor. But, in actuality, their attitudes and their work were precise and exact, and the crew, as always, knew that there was no room for error in the science of rocketry. The mission would likely fail if the smallest thing were out of place - - one wrong setting on an instrument or a left-over part on the work bench after the panel was closed. The recovery system, the scientific package, the data acquisition system, telemetry, trajectory, and electronic systems all had to operate flawlessly, to say nothing of the solid rocket motor in the Orion and its explosive ignition on the pad scheduled for tomorrow.

At 5.2 meters in length and about as big around as a telephone pole, the American-made Orion was relatively small as sub-orbital rockets go, but it was one of the dozen workhorses in the sounding rocket business. With its long nose that tapered down to a needle point and tail fins to keep it stable during flight, the Orion had that classic rocket silhouette of the early days when you knew that something that looked good would fly well. And the Orion flew well, shooting payloads of 100 to 200 kilograms into space, sometimes as high as 100 kilometers. Of course, Anders and his co-workers knew there was nothing here to even remotely approach the

glamor of a manned launch or even an orbital flight. But they still had all of the excitement, nervous worry, and the great sense of accomplishment of getting an important payload off the earth, through the atmosphere, and into space. And there was really no other place in the world quite like ESRANGE, where a rocket could be shot right into the northern lights to sample their mysteries or, in the case of tomorrow's launch, to collect scientific data in and just above the atmosphere where a satellite could not fly.

Were it not for the pair of geodesic radar domes and nearby launch pads, ESRANGE could be mistaken easily for a small, although remote, college campus nestled in the forested hills of northern Sweden. The closest town was Kiruna, 50 kilometers to the southwest. Although the townsfolk and native Laplanders of this sparsely populated corner of the world were pleased to have ESRANGE nearby and always treated the scientists and technicians well, there were still only two real sources of entertainment for the ESRANGE personnel living at this extreme latitude: watching the spectacular auroral displays in the cold winter days and witnessing a countdown and fiery rocket launch in this pristine faraway setting with a pastel northern sky as a backdrop. There was nothing like it in all the world, and tomorrow's launch into the ozone in the dead of a dark winter above the arctic circle was going to be spectacular in all respects.

It was always a little chilly in the long rocket assembly

hall at ESRANGE. The building was heated, but invariably felt cold, and the fact that the technicians ended up standing around for much of the day on the cement floor out in the assembly area did not help. The standard attire among this well-educated and relaxed-looking group was a wool sweater, warm slacks, and a pair of comfortable shoes.

The building itself was over 50 meters in length and rather narrow. And, as on most projects, the rocket being prepared for launch sat horizontally above the floor, its tail backed up near one end of the building and its probing needle-like nose pointed toward the largely empty space down the hall. There were three fins at the tail. They were nice big fins that arched back to complement the classic long lines of the rocket body and keep her on a straight and spin-stabilized course once her sturdy motor was ignited. She looked good just lying there on her side in the stand, as if she was already in motion, streaking toward space.

Today the one-stage Orion held a hefty 270 kilograms of solid fuel and oxidizer propellent that would burn smoothly but fiercely once ignited. The generated hot gases would explode out of the combustion chamber, through the nozzle at tremendous speed and pressure to generate forward thrust. Every action has an equal and opposite reaction; the hot gases shooting out the nozzle would be the action, and the forward pressure, or thrust, would be the reaction. If there were sufficient thrust to overcome the rocket's mass and the force of gravity, the Orion would blast forward in whatever direction she happened to be pointed. With nearly 600 pounds of highly explosive chemicals in the combustion chamber, the motor would generate more than enough thrust to accelerate the Orion off the pad. It was a tried and true American powerplant, proven through years of service throughout the western world.

Anders and his associates in the assembly building were

well aware of the dangers of working around rocket propellents and took seemingly every precaution imaginable. There were certainly no open flames nearby, there was nothing overhead that could fall onto the rocket, and the Orion itself was secure on its flatbed stand. The procedures had been perfected over the years and through hundreds of launches, and every step had been examined by the safety personnel and review boards. Nothing was taken lightly, even though they had accumulated a flawless record and knew practically every procedure by heart. Bengt Degerbjorck, ESRANGE's security officer, knew how important it was to do everything right: "Testing these rockets is equivalent to working in a gunpowder factory, and all the technicians are very experienced." "Safety rules are rigorous at ESRANGE," said Bengt Degerbjorck, and he knew what he was talking about.

Outside the assembly hall it was dark, not pitch-black, but the usual dark pink twilight that lingered throughout the day this time of year at this extreme latitude near the pole. The light inside the assembly hall was comfortably bright, and it bounced off the cream walls and cement floor and illuminated everything that needed to be seen around and inside the Orion. Anders and the crew had worked their way throughout almost all of the rocket's systems, beginning with the payload and on down towards the motor. They were now about to test the motor's igniter prior to closing up and calling it a day.

On the stand next to the rocket was the small hand-held test gauge, an Alinco model 101-5CFG, powered by a 1.3 volt battery about the size of what might be found in a common household flashlight. The tester would put a very weak charge to the rocket's igniter to check that everything was connected and working properly. The appropriate circuits on the Orion had been identified, and the wires had been run from the tester sitting on the stand next to the rocket over to the motor. Anders, the technician from Germany, and another Swede stood near the base of the Orion finishing their procedures and Bror Torneus, the other Swedish member of the launch team, had moved up beyond the rocket's nose. Anders picked up the tester in his hand to turn it on and route the low-level current through the igniter on the rocket motor.

What he did not know and could not know was that their string of 350 successful launches and decades-long safety record was going to come to an end during the next few seconds. Three men were about to be seriously injured, and another was going to die.

On any one-shot enterprise, especially one involving the launch of a very expensive rocket into space, it seems prudent to replace the batteries with fresh ones. A testing technician wants to be assured of putting just the precise amount of current through his equipment after all, and one of the easiest and cheapest ways to do that is to install new batteries, which is exactly what had been done to all of the test equipment in the days leading up to the launch. And the Alinco model 101-5CFG tester, a common device that had been used for years, especially in the U.S., had had its battery replaced in accordance with the requirement.

However, the Alinco model 101-5CFG tester was a bit unusual when it came to its battery and the slot behind the little door where it fit into place. Inside the battery compartment was not just a battery, but a plastic battery holder. And hidden inside the plastic battery holder was a tiny voltage reduction circuit that lowered the battery's 1.3 volts down to something just above a trickle, to a voltage that would run through the igniter's circuitry to test it out, but certainly not set off the rocket motor and its nearly 600 pounds of high-powered propellent.

But whoever had replaced the battery had removed the battery holder along with the old battery, and in its place slipped a new battery into the compartment without the holder. It was easy enough to have done. The new battery fit in the compartment just fine by itself; the small plastic battery holder, with no labeling or any other clear evidence to show that it contained the all-important voltage-reduction circuit, was left on a workbench. The 1.3 volt current of the battery would now flow out of the tester and into the igniter of the 5.2 meter rocket on the horizontal test stand in the assembly hall at ESRANGE above the arctic circle.

Anders flipped the switch on the test meter in his hand, anticipating the innocuous blip on the meter display as the small current ran though the igniter's circuitry on the rocket. It was only a fraction of a second until the potent solid fuel ignited within the combustion chamber and shrieked out the nozzle in a deafening blast of sound and flame, and only a fraction of a second later that the Orion rocket accelerated straight off its horizontal test stand in an eruption that blew the roof off the assembly hall. Still well ahead of the white fire and smoke, at least for the first split second, was Bror Torneus, who must have turned around to see what in the world was going on just as the long pencil nose of the Orion, followed by the knife-edge stabilizing fins, cut through him, then rocketed on through the far door at the other end of the hall, through one wall and out the other side of another big building at ESRANGE, and then into a snow-covered hillside where the explosion lit up the dark daytime sky above the arctic circle in northern Sweden.

REFERENCES AND NOTES

Pedersen, A., Colson, P., Meiner, R., Menardi, A. Sanderson, T. R. and Englund, P. (1997). Thirty years of sounding rockets - reflections following a reunion at ESRANGE, *ESA World Wide Web Site Link.*

Space rocket ignites while on flatbed, killing technician (1993). *New York Times,* March 1, A, 9:3.

Swedish launch site accident caused by test equipment (1993). *Space News,* March 8-14, 5.

Swedish Space Corporation, Esrange Satellite Station, Description (1997). *ESRANGE World Wide Web Site.*

A pseudonym has been used for the character of Anders Svanson. Bror Torneus, 44 years of age, died from his injuries.

ABOUT THE AUTHOR

Steven M. Casey, Ph.D., is President and Principal Scientist at Ergonomic Systems Design, Inc., a human factors research and design firm in Santa Barbara, California. He received his formal education in psychology and engineering from the University of California and North Carolina State University. His work has covered a range of products, systems, and settings, including industrial control centers, nuclear power plants, oil-field machinery, materials handling vehicles, portable electronic devices and computers, medical and environmental test equipment, automobile controls, agricultural and construction machinery, entertainment operations, and recreation and retail facilities. He travels extensively in his work, and has completed projects for clients in the United States, Canada, the United Kingdom, France, Germany, Belgium, Denmark, Finland, and Japan. The objective of his work is to help make things, especially things involving advanced technology, easier and safer for people to use. In 1987 he was presented the Alexander C. Williams, Jr. Award for outstanding human factors design contributions.